Polymerization Reactions
and New Polymers

Polymerization Reactions and New Polymers

Norbert A. J. Platzer, *Editor*

A symposium co-sponsored by
the Division of Industrial
and Engineering Chemistry
and the Division of Polymer
Chemistry at the 163rd
Meeting of the American
Chemical Society, Boston,
Mass., April 10–14, 1972.

ADVANCES IN CHEMISTRY SERIES **129**

AMERICAN CHEMICAL SOCIETY

WASHINGTON, D. C.   1973

ADCSAJ 129 1-287

Library of Congress Catalog Card 73-91734

ISBN 8412-0189-7

PRINTED IN THE UNITED STATES OF AMERICA

Advances in Chemistry Series

Robert F. Gould, *Editor*

FOREWORD

ADVANCES IN CHEMISTRY SERIES was founded in 1949 by the American Chemical Society as an outlet for symposia and collections of data in special areas of topical interest that could not be accommodated in the Society's journals. It provides a medium for symposia that would otherwise be fragmented, their papers distributed among several journals or not published at all. Papers are refereed critically according to ACS editorial standards and receive the careful attention and processing characteristic of ACS publications. Papers published in ADVANCES IN CHEMISTRY SERIES are original contributions not published elsewhere in whole or major part and include reports of research as well as reviews since symposia may embrace both types of presentation.

CONTENTS

PREFACE

Expansion of plastics into new and replacement markets promises a high rate of growth. Twenty five years ago suitable applications had to be found for the synthetic polymers manufactured at that time. Today's inventors visualize the need for certain materials. They prepare these materials and develop a feasible manufacturing process since macromolecules can be designed for a specific end use. Synthetic polymers can be made with flexible or stiff chains and tailored in length.

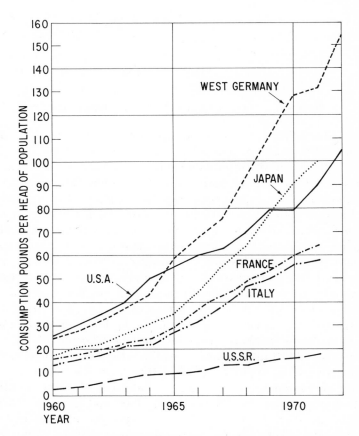

Figure 1. Plastics consumption per head of population in different countries

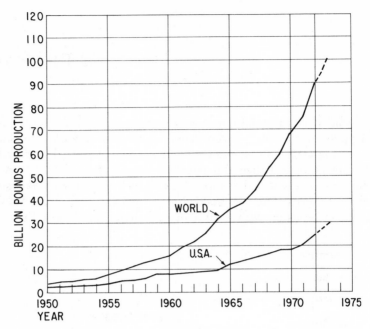

Figure 2.   World and domestic production of plastics

They can be polymerized in linear structure for fiber-forming purposes; branched with polar or nonpolar side chains for films, sheeting, or molded articles; or made crosslinkable as rubber, surface coatings or thermosets. They can be made as homopolymers or as random, graft, alternating, or block copolymers; mixed with plasticizers or other additives; blended with other synthetic resins; toughened with elastomers or reinforced with carbon black, minerals, or glass fibers. Specifications are given by the building, automotive, marine, aircraft, textile, furniture, and packaging industries. According to their demands, synthetic polymers are being produced of high strength, toughness, heat resistance, light weight, transparency, non-flammability, and low electrical and/or thermal conductivity. As specified, they can be made resistant to acids, solvents, and the environment, or they can be photodegradable. In surgical medicine, synthetic polymers must be biocompatible for use as implants, for substituting tissue, blood vessels, or part of organs.

The diversified applications of plastics have resulted in an increasing rate of consumption, as Figure 1 indicates. In this country consumption per person was 26 pounds in 1960 and grew to 105 pounds in 1972. In West Germany, Denmark, and Japan, consumption has exceeded ours. There seems to be no reason why the United States could not reach these higher levels.

Figure 2 compares the growth of U.S. domestic with total world production since 1950. It shows that the United States has reached 24.5 billion pounds in 1972 and will exceed the 30 billion pounds mark by 1974. Total world production was only 3 billion pounds in 1950, 16 billion pounds in 1960, and will reach 100 billion pounds in 1973 or 1974.

The demand for an enormous variety of plastics—spanning structural to medical applications, from gels and soft rubbers to rigid containers—is growing constantly. It is expressed in the diversity of chapters in this book which contains papers on new polymers and novel polymerization reactions. These papers were presented at the sixth symposium on polymerization and polycondensation processes held by the Division of Industrial and Engineering Chemistry and the Division of Polymer Chemistry during the American Chemical Society Meeting in Boston, April 9-14, 1972. At that symposium, papers were also given on kinetics and technology of commercial polymers and their manufacturing processes. The latter are collected in the preceding volume, ADVANCE IN CHEMISTRY SERIES No. 128.

New Polymers

Plastic fabricators and customers generally don't ask for chemical compositions but look at a material as a bundle of properties. Although we are unable to incorporate all desired properties into a single product, we can combine several of them in special or more universal polymers.

Elastomers. Polycyanoprene represents a new synthetic rubber. In mechanical properties it is similar to the commercial polychloroprene. However, because of the presence of a nitrile rather than a chlorine group, it has better oil resistance. Its monomer synthesis, polymerization, and properties are reported in Chapter 1 by E. Mueller, R. Mayer-Mader, and K. Dinges. Terpolymers of tetrafluorethylene, perfluoro(methyl vinyl ether), and certain cure site monomers are other inert elastomers reported by G. H. Kalb, A. L. Barney, R. W. Quarles, and A. A. Khan. Hydrogenation of styrene-butadiene block copolymers has led to novel transparent polymers of high heat stability superior to the original materials. This reaction using soluble transition metal catalysts was investigated by J. F. Pendleton, D. F. Hoeg, and E. P. Goldberg. Segmented copolymers of polyethers and polyesters are another family of new elastomers introduced at the symposium. The polyether provides the soft segment and the polyester the rigid segment. W. K. Witsiepe describes the polycondensation of polytetramethylene ether glycol 1,4-butanediol with dimethyl terephthalate and the different physical properties which result by varying the components.

Heat Resistant Polymers. Perfluoroalkylene ethers are characterized by high chemical and thermal stability. Their oligomers with dicarboxylic acid were used as building blocks for isocyanurate polymers and polyimides by J. A. Webster, J. M. Butler, and T. J. Morrow. Polyphenylene sulfides are a unique family of heat and chemical resistant materials. Until recently they had to be prepared by heating cuprous *p*-bromothiophenoxides. A new economical process for their commercial manufacture by the reaction of *p*-dichlorobenzene and sodium sulfide in a polar solvent was discovered at Phillips Petroleum Co. A 6-million pounds/year plant for producing polyphenylene sulfides under the trade name Rytex started up in December 1972. The products may be used either as high melting molding powders or protective coatings. H. W. Hill Jr. and J. T. Edmonds, Jr. describe the properties of the coating grade. Within the last few years, several cyclic bis(arylene tetrasulfides) have been synthesized. They can be polymerized in a free radical process to stable polymers applicable as tire-cord adhesive or metal-to-metal binder, as presented by N. A. Hiatt. Poly(thiol esters) can be prepared analogously to polyesters. H. G. Buehrer and H. G. Elias show in their chapter that the melting points of the poly(thiol esters) are 30°–70°C higher than those of the corresponding polyesters but lower than those of the corresponding polyamides. Poly(thiol esters) appear to be well suited for fabrication into fibers, film, or moldings. A practical route to new linear aromatic polysulfones has been developed by R. J. Cornell. He mixed bischlorophenyl sulfone with the disodium salt of bisphenol obtained from diisopropylbenzene and phenol. The resulting thermoplastic is transparent and suitable for injection molding.

Novel Polymerization Reactions

Polymerization by Ring Opening. A novel technique of making polyolefins by ring opening and simultaneous polymerization starting from small cyclic hydrocarbons is revealed by C. P. Pinazzi and his co-workers. Substituted cyclopropanes were polymerized by cationic and Ziegler-Natta catalysts. Pivalolactone can be polymerized anionically to yield a linear polyester exhibiting properties suitable for textile fiber or structural plastic. Its two commercial processes by mass and slurry polymerization are discussed by N. R. Mayne.

Alternating and Block Copolymerization. Random and graft copolymerization are well established processes. More recent are alternating and block copolymerization. Alternating copolymerization of α-olefins with fluoroacetone by γ-rays is dealt, with by Y. Tabata, W. Ito, Y. Yamamoto, and K. Oshima. Alternating copolymers of styrene with maleic anhydride, methyl methacrylate, and acrylonitrile have been prepared by N. G. Gaylord. He describes grafting these alternating copolymers on polystyrene and poly(butyl acrylate). New block copolymers of styrene with maleic anhydride are presented by R. B. Seymour and co-workers, and with dimethyl siloxane by J. C. Saam, A. H. Ward, and F. W. Fearon. Cationic block copolymerization of styrene with cyclic ethers, such as tetrahydrofuran, initiated by macromolecular dioxolenium salts is the subject of Y. Yamashita's chapter.

Photopolymerization. Photopolymerization in the organic crystallin state has become a promising method for preparing new polymers. G. Wegner reviews the kinetics of solid-state photoreactions, the geometric molecular structure, and the lattice controlled photoreactivity (topochemistry). He reports on the four-center type photopolymerization and on the topochemical polymerization of monomers with conjugated triple bonds. A commercial photopolymerization process for insulating magnet wire was developed by E. D. Feit. The wire is passed first through an applicator where it is coated with an urethane modified acrylate and then through a reactor where polymerization takes place under ultraviolet light.

Conclusion

The last chapter closes the circle with the previous volume which dealt with commercial processes and products. Together both volumes represent an overall cross-section of the improvements of existing processes and modifications of products as well as of new polymers and novel polymerization reactions, candidates of the future polymer industry. Polymer science and technology are young and diversifying, and the polymer industry will maintain its growth and expand into new fields of applications.

NORBERT A. J. PLATZER

Longmeadow, Mass.
December 1972

Production and Properties of 2-Cyano-1,3-butadiene Homo- and Copolymers

E. MÜLLER, R. MAYER-MADER, and K. DINGES

Bayer, Leverkusen, West Germany

2-Cyano-1,3-butadiene was synthesized by pyrolysis of 1-cyano-1-cyclohexene and characterized by physical data. Soluble homopolymers and copolymers with butadiene, isoprene, styrene, and chloroprene have been prepared by solution polymerization under free-radical, anionic, and Ziegler-Natta initiation. Radical homopolymerization in ethers or toluene/ether mixtures yielded thermoplastic polymers consisting mainly of 1,4- and 3,4-adducts in a ratio of 3:1. The polymers show glass stages of about 10°C. Anionic initiation below 0°C with lithium alkyls, metal phosphide, or metal amide catalysts gave thermoplastic products with 57–70% of 1,2-linkages and glass stages of about 80°C. Initiation with metal acetyl acetonates and organoaluminum compounds resulted in polymers that in large part correspond to the radical-produced polymers. Block polymerization was carried out with lithium alkyl compounds, as well as with alkali amides or alkali phosphides. Cyanoprene polymerized onto polymer ions formed of styrene, butadiene, or isoprene; the reverse reaction does not occur.

Polychloroprene has a number of desirable properties such as weather and ozone resistance, flame retardance, and medium oil resistance. Its swelling behavior toward aromatic oils, however, is moderate, and inferior to that of butadiene/acrylonitrile copolymers. During work on the synthesis by adding HCN to vinylacetylene. Instead, HCN addition tion and copolymerization behavior of comonomers suitable for this purpose was studied generally. In view of its constitution and the presence of a nitrile group, cyanoprene (2-cyano-1,3-butadiene) appeared suitable and, owing to little reference in the literature, especially interesting.

Monomer Synthesis

2-Cyanobutadiene is not accessible along the lines of chloroprene synthesis by adding HCN to vinylacetylene. Instead, HCN addition results in 1-cyano-1,3-butadiene.

$$2\ HC \equiv CH \longrightarrow H_2C = CH - C \equiv CH$$

$$H_2C=C-CH=CH_2 \quad\quad HC=CH-CH=CH_2$$
$$\quad\quad\ |\qquad\qquad\qquad\ |$$
$$\quad\quad Cl\qquad\qquad\qquad CN$$

a) HCl   b) HCN

There are three processes described in the literature for synthesizing cyanoprene (1-5).

a)
$$\begin{array}{c} CH_3 \\ | \\ C=O \\ | \\ CH \\ \| \\ CH_2 \end{array} \xrightarrow{HCN} \begin{array}{c} CH_3 \\ | \\ C{<}^{OH}_{CN} \\ | \\ CH \\ \| \\ CH_2 \end{array} \xrightarrow{(Ac)_2O} \begin{array}{c} CH_3 \\ | \\ C{<}^{OCOCH_3}_{CN} \\ | \\ CH \\ \| \\ CH_2 \end{array}$$

pyrolysis $\downarrow$

$$H_2C = C - CH = CH_2$$
$$\qquad\quad |$$
$$\qquad\quad CN$$

b) $2\ H_2C = CHCN \longrightarrow \square{<}^{CN}_{CN} \xrightarrow{pyrolysis} H_2C = C - CH = CH_2$
$$\qquad\qquad\qquad\qquad\qquad\qquad\qquad\qquad\qquad |$$
$$\qquad\qquad\qquad\qquad\qquad\qquad\qquad\qquad\qquad CN$$

c) $H_3C - \begin{array}{c} CH_2 \\ \| \\ C \\ | \\ CH \\ \| \\ CH_2 \end{array} \xrightarrow{O_2/NH_3} \begin{array}{c} CH_2 \\ \| \\ C - CN \\ | \\ CH \\ \| \\ CH_2 \end{array}$

In Process a, vinyl methyl ketone reacts with hydrocyanic acid to give cyanohydrin, which is either dehydrated by spraying with phosphoric acid at 540°C (yield 74%) or, after acylation with acetic anhydride, pyrolized at 400°-550°C. In Process b, acrylonitrile is dimerized by UV irradiation to give 1,2-dicyanocyclobutane; 8.5% of the acrylonitrile is converted after 84 hours. The cyclobutane derivative reacts by catalytic splitting at 455°C to give 2-cyanobutadiene (conversion 20%). Process c involves ammonoxidation of isoprene. One per cent by volume of isoprene reacts with ammonia, steam, and air at 425°C on zeolite catalysts containing copper (conversion 30%, yield 60%). The process used by us (6) is based on the general route of synthesizing 1,3-dienes by retro-diene splitting of cyclohexene derivatives.

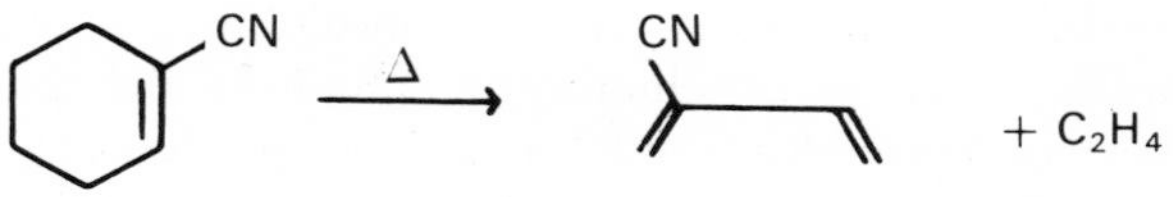

The synthesis goes from cyclohexanone *via* cyanohydrin to 1-cyano-1-cyclohexene, which is split by pyrolysis.

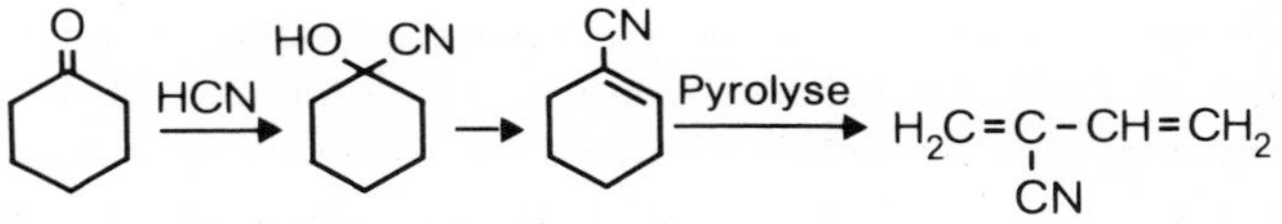

The yield of 2-cyanobutadiene is 75%.

Properties of *2-Cyano-1,3-butadiene*

2-Cyanobutadiene is liquid at room temperature. Its boiling point at 760 torr cannot be measured directly but can be extrapolated from the vapor pressure curve: 90°C. At 4 torr, its boiling point is 0°C. Its density at $-30°C$ is 0.89. The compound has a strong tendency to polymerize and dimerize. Polymerization, but not dimerization, can be prevented by adding phenothiazine. Dimerization takes place quickly at room temperature, various kinds of dimers being possible. For example, 1,4-dicyano-4-vinylcyclo-1-hexene forms in about 86% yield (7). Stabilization of 2-cyanobutadiene against dimerization can be achieved by storage at low temperatures and by dilution with benzene, for example.

Homopolymerization of *2-Cyano-1,3-butadiene*

Tanaka (8) reported polymerization experiments with benzoyl peroxide. He observed the formation of insoluble polymers in addition to dimer. Wei (9) carried out polymerizations with butyllithium and organoaluminum compounds and obtained amorphous and partly crystalline products with thermoplastic properties; the materials melted at 157°C and decomposed at 336°C. IR spectroscopy showed, a 1,4-trans structure.

In our work, the homo- and copolymerization of 2-cyanobutadiene were studied from the point of view of the initiation system (radical, ionic, and coordinate), and the microstructure and properties of the products obtained were determined.

Radical-Initiated Homopolymerization. When this homopolymerization is carried out with benzoyl peroxides or other radical formers in a manner analogous to emulsion polymerization of chloroprene, highly crosslinked polymers are formed. They are insoluble in organic solvents such as toluene, benzene, or chloroform. Radical polymerization in toluene, benzene, or hexane leads only to insoluble products.

In assessing these results, it should be remembered that the temperatures in most cases were above 0°C so that dimer formation has to be taken into account. On the one hand, dimerization reduces the yield and, on the other hand, it influences the course of polymerization through a regulating effect. Since we wished to obtain soluble polymers, we had to look for different routes. In doing so, we found that soluble polymers can be obtained when the reaction is carried out in polar solvents such as ethers or in inert solvents such as toluene or hexane in the presence of cocatalysts—ethers, phosphines, amines, or organoaluminum compounds. The polymers obtained are thermoplastic and consist mainly of 1,4-and 3,4-adducts in a 3:1 ratio.

Besides the 1,4- and 3,4-adducts, cyclic structures are also present. Four representative radical polymerizations of cyanoprene are compared in Table I. Differential thermal anlysis shows, for the material made from toluene, a glass stage of about 10°C and a melting range above 90°C. The material has medium crystallinity while the material made in THF is largely amorphous.

The disadvantage of the soluble polycycanoprenes obtained by a radical reaction was that they have molecular weights below 5000, and the yields were not always good. Therefore, it was necessary to look for different catalyst systems.

Anionically Initiated Polymerization. The disadvantages of radical polymerization of cyanoprene result from the operating conditions (temperatures); too many side reactions, chain-terminating reactions, and consecutive reactions occur. Because of this and the dimerization tendency of cyanoprene, catalysts had to be found that could fulfill two contradictory requirements. They should be so reactive that it would be possible to work at temperatures that exclude dimerizations as com-

Table I. Radical Polymerizations of Cyanoprene

Solvent	Monomer (mole/l)	Catalyst (mmole/l)	Temperature (°C)	Reaction time (h)	Yield (%)	Soluble in [e]	N Analysis[f] (%)
Emulsion in Water	10	FAS 60[a]	50	5	50	—	17.1
Toluene	0.29	AIBN 14.5[b]	50	24	50	—	17.9
THF	0.29	AIBN 14.5	50	24	50	DMF[d]	16.9
Toluene	0.8	(AIBN + TEA[c]) 8.7	50	5	20	DMF	16.9

[a] FAS = formamidine sulfinic acid
[b] AIBN = azodiisobutyronitrile
[c] TEA = triethylaluminum
[d] DMF = dimethylformamide
[e] Theoretical nitrogen value = 17.7
[f] The molecular weights of the soluble products were from 1000 to 5000.

peting reactions, and they should be slow enough to prevent crosslinking and cyclization reactions at low temperatures.

In view of these preconditions, it was obvious to use anionic initiators. With the lithium alkyls used by Wei—for example, butyllithium or other organic alkali compounds—only insoluble thermoplastic products are obtained, even at −80°C in both THF and toluene because these initiators are too reactive. The reactions that take place include attack by the metallo-organic compound and the already metallized cyanoprene (or the metallized polymer) on the nitrile groups of both monomer and polymer.

Anionic catalysts of the type shown here proved to be suitable because they react comparatively smoothly and, unlike organo-lithium compounds, they do not attack the CN group.

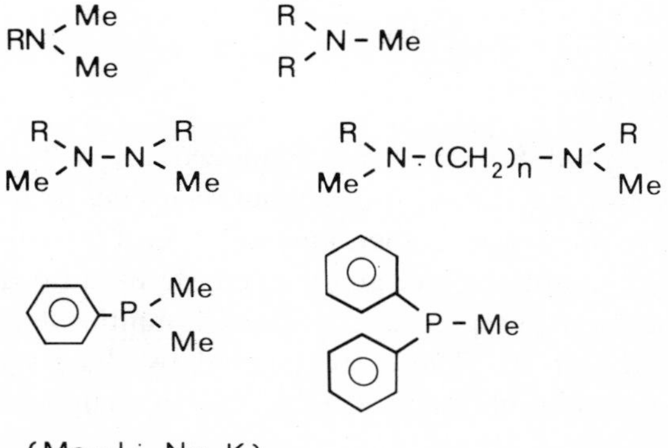

(Me = Li, Na, K)

In carrying out these experiments, a solution of the monomer is added dropwise into a mixture of catalysts and solvents. The catalyst is destroyed, and the polymer obtained precipitated with methanol. The products precipitated from methanol are plastic, owing to the plasticizing effect of the solvent. They swell on drying in vaccum, and finally yield thermoplastic, porous solids. Depending on the experimental conditions and the catalyst used and its quantity, the molecular weights of the products are about 10,000.

The product studied was produced in THF with a diphenylphosphine–lithium catalyst; it had a molecular weight of 8300. After shear modulus plotting over temperature, a glass stage of 81°C, a modulus of elasticity of 32,000 kg/cm^2 (polystyrene~30,000), and a flexural strength of 661 kg/cm^2 (polystyrene~1000) were found. The glass temperature was 20°C lower than that of polystyrene, but the polymer is more resistant to swelling by aromatics.

Table II. Cyanoprene Homopolymers Produced by Different Catalyst Systems [a]

Catalyst	Yield (%)	$\overline{M}_n$	N Analysis	Adducts (%) 1,2	1,4	3,4	Cyclic Structures (%)
ϕ_2 PK	96		16				
ϕ_2 PK + N(C$_2$H$_5$)$_3$	34	8,240	16.5	62		<1.4	Present
ϕ_2 PLi	79	8,240	16.6	70		30	
ϕ PK$_2$	91	>10,000	16.3	57.6	29	4.4	9
ϕ_2 NLi	93		16.8	66.9	17.7	0.4	15
(n-C$_3$H$_7$)$_2$NLi	53		16.6	70	17.8	1.3	10.9
n-C$_4$H$_9$Li [b]	90						
n-C$_4$H$_9$Li + P(n − C$_4$H$_9$)$_3$ [c]	93	2,310	15.5	64.1	28.1	7.0	0.7
ϕ Na [b]	59		16.2				

[a] Temperature: −78°C; solvent: THF
[b] Insoluble polymers
[c] Toluene as the solvent

The solvent used for most of polymerizations is THF or dioxane. However, it is also possible to use toluene, benzene, or hexane with an addition of polar substances (ether, amines). In those cases, it is necessary to use mild conditions and gentle catalysts because if the dissociation of the amides of phosphides is weakened, multicenter processes, for example, have an effect. These are favored by the free electron pair on the nitrogen or phosphorus, and they lead to chain-terminating reactions.

Regarding the structure of the cyanoprene homopolymers (Table II), a trend becomes apparent when using the different catalyst systems. In a sample polymerized with diphenylphosphine–lithium, 70% 1,2-linkages and about 30% 1,4-linkages were found, but the presence of traces of 3,4-adducts could not be excluded. There was no indication of cyclic structures although with stronger diphenylphosphine–potassium cyclic compounds were found. About 62% were 1,2-linkages while the remainder consisted of about equal parts of 1,4- and 3,4-adducts, the cyclic proportions included.

When triethylamine was added to this catalyst system, the proportion of 1,2-linkages remained about equal, but the proportion of 1,4-structures grew compared with the polymer made *via* pure diphenylphosphine potassium. Polycyanoprenes produced with the more reactive metal amide catalysts showed greater proportions of cyclic structures (11-15%).

Thermal study of the anionically polymerized homopolymers showed glass-transition temperatures from 40° to 120°C. The samples partly changed on heating; on second heating, the second-order transition points were often higher.

Rearragements, cyclizations, and the like occur, manifesting themselves in an exothermic reaction above 200°C. Exothermic decomposition occurs above 350°C. Considering the polymerization mechanism, there are at first two possibilities as shown below.

1 st mechanism

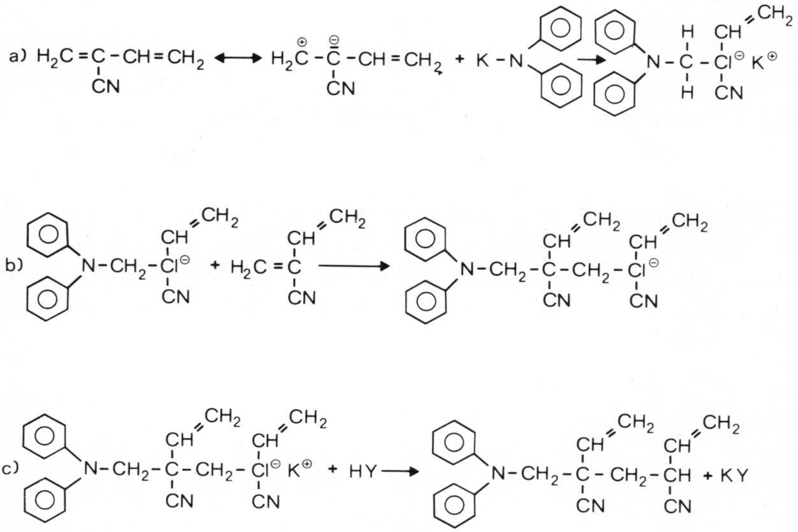

2nd mechanism

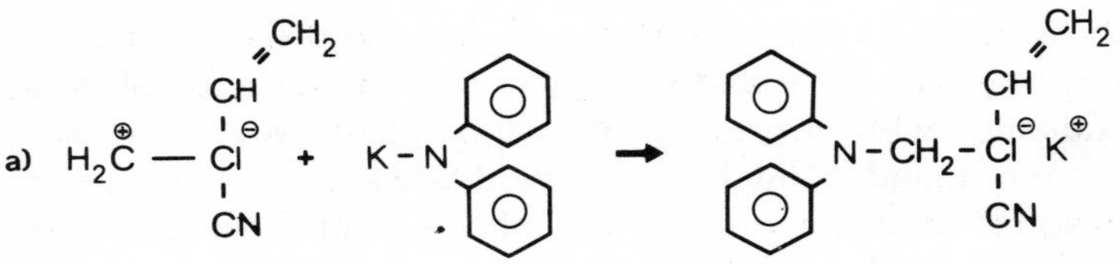

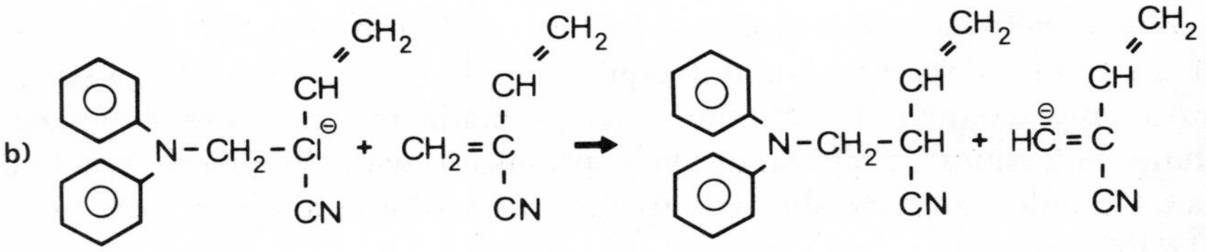

a) H$_2$C$\overset{\oplus}{}$ — $\overset{\ominus}{Cl}$ (with CH, CH$_2$, CN substituents) + K—N(carbazole) → N—CH$_2$—$\overset{\ominus}{Cl}$ K$^{\oplus}$ (with CH, CH$_2$, CN)

b) (carbazole)N—CH$_2$—$\overset{\ominus}{Cl}$ (with CH, CH$_2$, CN) + CH$_2$=C (with CH, CH$_2$, CN) → (carbazole)N—CH$_2$—CH (with CH, CH$_2$, CN) + H$\overset{\ominus}{C}$=C (with CH, CH$_2$, CN)

The broad molecular-weight distribution, the low or high molecular weights (depending on the activity of the catalyst), and the fact that added butadiene or isoprene does not polymerize favor the second mechanism. Nevertheless, under certain conditions, a "living polymer" cannot be excluded, as shown by segment polymerization experiments and kinetic studies at low temperatures ($-78°C$).

Cyanoprene can also be polymerized with tributylphosphine or other trialkylphosphines but only in toluene, benzene, etc., and, not in ethers. The mechanism is probably similar to that with amides or macroions in a "living polymer" mechanism, or, again, transfers would have to be assumed. The former may be assumed this way:

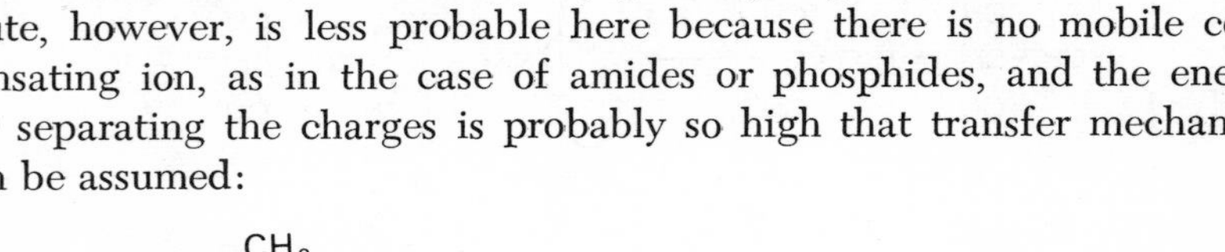

R$_3$P + CH$_2$=C (CH$_2$, CH, CN) → R$_3\overset{\oplus}{P}$—CH$_2$—$\overset{\ominus}{Cl}$ (CH$_2$, CH, CN) $\xrightarrow{\text{cyanoprene}}$ R$_3\overset{\oplus}{P}$—[CH$_2$—C (CH$_2$, CH, CN)]$_n$—CH$_2$—$\overset{\ominus}{Cl}$ (CH$_2$, CH, CN)

This mechanism results in a steadily growing amphoteric macroion. This route, however, is less probable here because there is no mobile compensating ion, as in the case of amides or phosphides, and the energy for separating the charges is probably so high that transfer mechanism can be assumed:

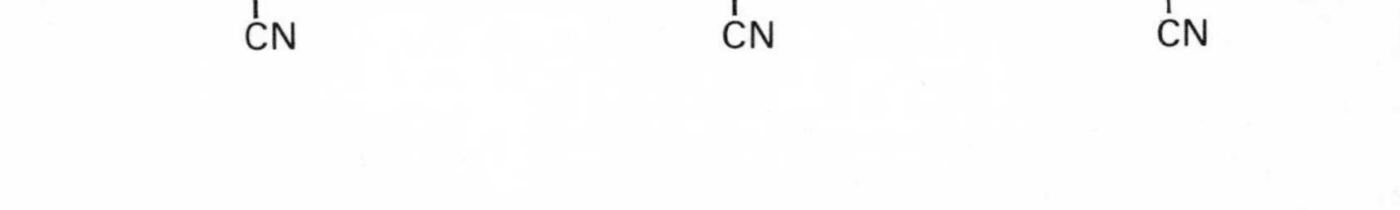

R$_3\overset{\oplus}{P}$—CH$_2$—$\overset{\ominus}{Cl}$ (CH$_2$, CH, CN) + Cyap → R$_3\overset{\oplus}{P}$—CH$_2$—CH—CH=CH$_2$ (CN) + $\overset{\ominus}{C}$H=C—CH=CH$_2$ (CN)

Only low-molecular-weight phosphonium molecules are formed.

As mentioned elsewhere in this paper, the polycyanoprenes anionically polymerized by us contain 57–70% of 1,2-linkages.

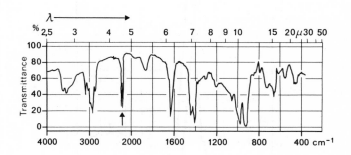

Correspondingly, the infrared spectrum shows two CN vibrations at 2210 and 2230 cm^{-1} (Figure 1). When the polycyanoprene is heated to 130°C for one hour, the band at 2230 cm^{-1} disappears (Figure 2).

Thermogravimetric analysis shows that splitting off the allyl-positioned cyanogen group as HCN cannot be involved here. In addition, the remaining cyanogen band increases. NMR investigations show that a change occurs at the vinyl-position protons. Since the molecular weight increases on heating, crosslinking reactions are very probable.

Figure 1.   Infrared spectrum of anionically polymerized cyanoprene

The spectroscopic data do not permit excluding cyclization reactions to a greater or lesser degree. Why the remaining cyanogen band increases must, however, still be considered unclarified.

Coordinate Homopolymerization.   When Ziegler-Natta catalysts of the type TiCl$_4$/alkylaluminum compounds are used, no polymerization occurred because the cyanoprene (like acrylonitrile for instance) reacts with the catalyst and destroys it. Polymerization occurs, however, when metal acetyl acetonates and organoaluminum compounds are used. For example, coordinate polymerization with a mixture of cobalt acetyl acetonate and ethylaluminum dichloride results in a polymer that corresponds mainly to the radical-produced polymer.

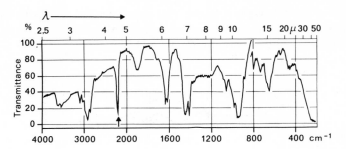

Figure 2. *Infrared spectrum of anionically polymerized after heating to 130°C for one hour*

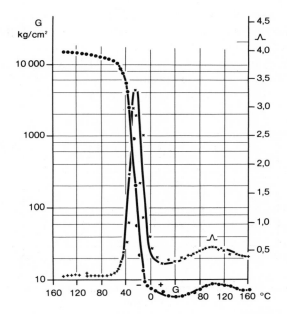

Figure 3. *Emulsion copolymer of chloroprene/cyanoprene (14% cyanoprene polymer); shear modulus plotted over temperature.*

Copolymerization of 2-Cyano-1,3-butadiene

There is only one hint in the literature, by Starkweather (*10*), that deals with copolymerization with butadiene. The polymers described there have poor tensile strength values and elongations.

Radical Copolymerization. We carried out an emulsion copolymerization with chloroprene at a chloroprene:cyanoprene ratio of 9:1. At 60% conversion, an insoluble copolymer with cyanoprene "lumped out" during the polymerization. The glass-transition temperature of −24°C showed that a genuine copolymer actually had been formed (Figure 3).

These copolymers have a strong odor, and discolor when exposed to light. The vulcanizates showed a strong tendency to scorch and showed strength and elongations below the level of comparable products. The corresponding copolymerization of isoprene–cyanoprene (9:1) gave, after vulcanization, crumbling products without any mechanical strength. The corresponding copolymerization in solution resulted in soluble products that have not yet been tested.

Anionic Copolymerization. For normal mixing-type copolymerization, the cyanoprene was added to the solvent containing the catalyst and the comonomer. Depending on the initial proportions, of monomers, products were obtained with varying cyanoprene proportions and correspondingly different properties.

For example, in one experiment, toluene, chloroprene, and tributylphosphine were placed in a vessel, and a cyanoprene-toluene solution was aded dropwise at 0°C. The result was a 60% yield of a thermoplastic product with 14.2% nitrogen, corresponding to a chloroprene proportion of 17%. Analogous experiments were carried out with monomers such as styrene and isoprene, and with other catalysts.

A further possibility consists of subjecting the catalyst to incipient polymerization with small quantities of styrene, isoprene, or butadiene (organic alkali compounds also can be used for this purpose), then adding dropwise a mixture of cyanoprene and selected monomers at low temperatures. If no incipient polymerization takes place, metalloorganic catalysts will only give a homopolymerization of the cyanoprene.

When block copolymerization was done, isoprene, styrene, or butadiene was prepolymerized at room temperature, and cyanoprene polymerized at low temperatures. Styrene gave thermoplastic polymers; butadiene, depending on the quantity ratio selected, gave thermoplastic to rubberlike polymer soluble in polar solvents such as DMF, but largely resistent to organic solvent such as toluene.

Acknowledgment

The authors thank G. Bayer, J. Thies, and D. Wendisch, whose spectroscopic work made possible the assignments of structures. We also thank G. Hentze for differential thermal analysis.

Literature Cited

1. Carter, S., Johnson, F. W., U.S. Patent **2,205,239**.
2. British Patent **482,300**.

3. Grasselli, R. K., Heights, G., Greene, J. L., Heights, W., Gray, N. R., U.S. Patent **3,347,902**.
4. Runge, I., Kache, R., British Patent **1,095,849**.
5. Hosaka, S., Wakamatsu, S., *Tetrahedron Lett.* (1968) (2) 219.
6. German Patent **622,800**.
7. Marvel, C. S., Brace, N. O., *J. Amer. Chem. Soc.* (1949) **71**, 37.
8. Tanaka, M., *Kogyo Kagaku Zasshi* (1958) **60**, 1367.
9. Wei, P. E., *J. Polym. Sci., P. A-1* (1969) **7**, 2305.
10. Starkweather, H. W., *Ind. Eng. Chem.* (1947) **39**, 210.

RECEIVED May 9, 1972.

Terpolymers of Tetrafluoroethylene, Perfluoro (Methyl Vinyl Ether), and Certain Cure Site Monomers

G. H. KALB, A. A. KHAN, R. W. QUARLES, and A. L. BARNEY

Research Laboratory, Elastomer Chemicals Department, E. I. du Pont de Nemours and Co., Inc., Wilmington, Del. 19898

Copolymers of tetrafluoroethylene, perfluoro(methyl vinyl ether), and a third monomer—selected from the group of perfluoro(4-cyanobutyl vinyl ether), perfluoro(4-carbomethoxbutyl vinyl ether), perfluoro(2-phenoxypropyl vinyl ether), or perfluoro(3-phenoxypropyl vinyl ether)—give vulcanizable, high performance elastomers. Syntheses of the cyano and the carbomethoxy compounds from perfluoroglutaryl chloride and the phenoxy compounds from pentafluorophenol are presented. Vulcanization methods for the three copolymer types include catalytic condensation of the nitrile, interchange of dibasic amines or esters with the methyl ester, and nucleophilic displacement of a fluorine atom on the phenoxy ring with a diamine or an aromatic bisnucleophile. A brief description of vulcanizate properties demonstrates the outstanding solvent and chemical resistance that may be obtained with these new elastomers.

A new high-performance elastomer, prepared from tetrafluoroethylene (TFE) and perfluoro(methyl vinyl ether) (PMVE), and characterized by outstanding resistance to chemical attack and excellent thermal stability, was reported recently by Barney *et al.* (*1*) of our laboratory. That paper described a rubbery perfluorinated dipolymer that could not be crosslinked using ordinary techniques because of its chemical inertness, and also a terpolymer in which an unspecified third monomer was used to introduce active crosslinking sites.

The dipolymer, which loses only 5 to 7% of its weight in an air-circulating oven at 316° C in seven days (and only a few per cent more a month at these conditions) is, however, a thermoplastic. It is apparent,

13

therefore, that a crosslinked structure is necessary to gain the mechanical strength at elevated temperatures required for a true high-performance elastomer.   This paper describes work aimed at: development of a suitable crosslink structure and various candidate third monomers that copolymerize with TFE and PMVE to introduce the necessary crosslinking sites; their preparation and copolymerization; crosslinking reactions involved with some of the monomers; and a few of the physical properties of the crosslinked ("vulcanized") compositions.

The dipolymer has many desirable properties, and, to maintain these properties, the requirements for the third monomer and the crosslinking reaction are quite stringent.   These requirements are:

(a) The cure site monomer must copolymerize readily without appreciable chain transfer (the polymerization is particularly susceptible to chain-transfer reactions), and at a rate such that reasonable and controlled amounts of crosslinking site are incorporated at an adequate spacing.

(b) The crosslinking functional group must not be affected by the preferred aqueous polymerization system.

(c) The crosslinking reaction must be such that the stock can be formed into desired shapes at elevated temperatures (by molding or other means), then be converted into a crosslinked structure.

(d) The crosslink should be comparable in thermal, chemical, and oxidation resistance to the backbone to maintain the outstanding properties of the dipolymer.

$$CF_2 = CF - O - (R_f) - X$$

where $X = -COOR, \ -CN, -O-C_6F_5$

and $(R_f) = $ PERFLUOROALKYL or
PERFLUOROALKYL ETHER GROUP

Figure 1.   Comonomer candidates—generalized structure

A survey of a variety of polymerizable groups indicated that in the preferred free-radical aqueous-emulsion polymerization system, a perfluorovinyl ether most nearly fulfills these requirements.   In general, the perfluorovinyl ethers copolymerize at a reasonable rate.   They change little in rate with variations in chain length or functionality in the perfluoroalkyl moiety so long as the type of functionality on the perfluoroalkyl group is consistent with requirements (a) and (b).   In addition, such ethers furnish thermally resistant structures in the polymer.

A number of monomers of the general type $CF_2{=}CF{-}O{-}R_fX$ (2, 3), where X is –COOR, –CN, and –OC_6F_5 (Figure 1) have been copolymerized with TFE and PMVE.   Their crosslinking reactions with difunctional vulcanization agents were carried out and the vulcanizate

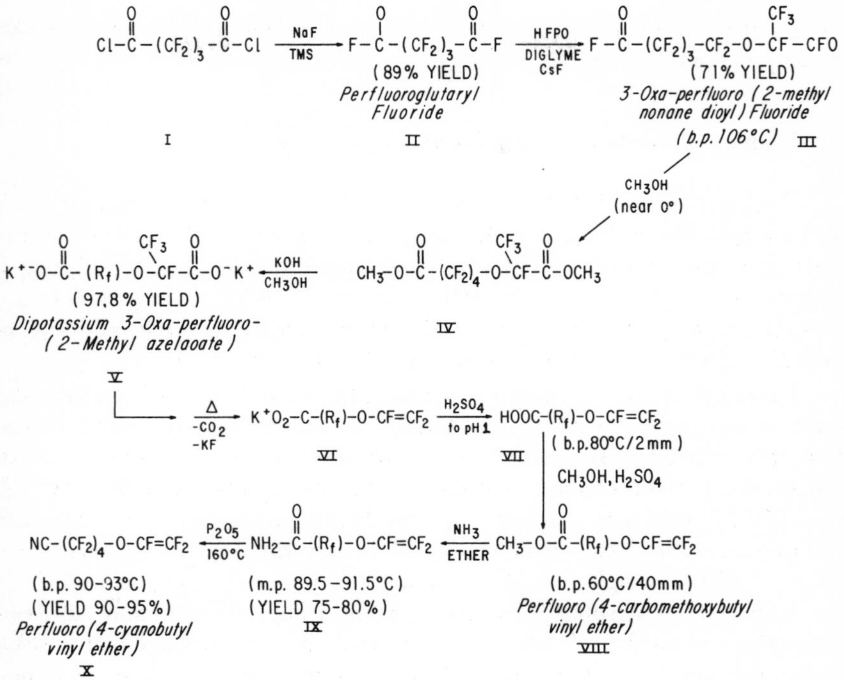

$$\underset{\text{I}}{Cl-\overset{O}{\overset{||}{C}}-(CF_2)_3-\overset{O}{\overset{||}{C}}-Cl} \xrightarrow[\text{TMS}]{\text{NaF}} \underset{\substack{\text{(89% YIELD)}\\ \text{Perfluoroglutaryl}\\ \text{Fluoride}\\ \text{II}}}{F-\overset{O}{\overset{||}{C}}-(CF_2)_3-\overset{O}{\overset{||}{C}}-F} \xrightarrow[\substack{\text{DIGLYME}\\ \text{CsF}}]{\text{HFPO}} \underset{\substack{\text{(71% YIELD)}\\ \text{3-Oxa-perfluoro (2-methyl}\\ \text{nonane dioyl) Fluoride}\\ \text{(b.p. 106°C)} \quad \text{III}}}{F-\overset{O}{\overset{||}{C}}-(CF_2)_3-CF_2-O-\overset{CF_3}{\overset{|}{C}F}-CFO}$$

CH$_3$OH
(near 0°)

$$\underset{\substack{\text{(97.8% YIELD)}\\ \text{Dipotassium 3-Oxa-perfluoro-}\\ \text{(2-Methyl azelaoate)}\\ \text{V}}}{K^{+-}O-\overset{O}{\overset{||}{C}}-(R_f)-O-\overset{CF_3}{\overset{|}{C}F}-\overset{O}{\overset{||}{C}}-O^-K^+} \xleftarrow[\text{CH}_3\text{OH}]{\text{KOH}} \underset{\text{IV}}{CH_3-O-\overset{O}{\overset{||}{C}}-(CF_2)_4-O-\overset{CF_3}{\overset{|}{C}F}-\overset{O}{\overset{||}{C}}-OCH_3}$$

$$\underset{\text{V}}{\big|} \xrightarrow[\substack{-CO_2\\ -KF}]{\Delta} \underset{\text{VI}}{K^+O_2-C-(R_f)-O-CF=CF_2} \xrightarrow[\text{to pH 1}]{H_2SO_4} \underset{\substack{\text{(b.p.80°C/2mm)}\\ \text{VII}}}{HOOC-(R_f)-O-CF=CF_2}$$

CH$_3$OH, H$_2$SO$_4$

$$\underset{\substack{\text{(b.p. 90-93°C)}\\ \text{(YIELD 90-95%)}\\ \text{Perfluoro(4-cyanobutyl}\\ \text{vinyl ether)}\\ \text{X}}}{NC-(CF_2)_4-O-CF=CF_2} \xleftarrow[\text{160°C}]{P_2O_5} \underset{\substack{\text{(m.p. 89.5-91.5°C)}\\ \text{(YIELD 75-80%)}\\ \text{IX}}}{\overset{O}{\overset{||}{NH_2-C}}-(R_f)-O-CF=CF_2} \xleftarrow[\text{ETHER}]{NH_3} \underset{\substack{\text{(b.p.60°C/40mm)}\\ \text{Perfluoro (4-carbomethoxybutyl}\\ \text{vinyl ether)}\\ \text{VIII}}}{CH_3-O-\overset{O}{\overset{||}{C}}-(R_f)-O-CF=CF_2}$$

Figure 2. Synthesis of perfluoro(4-cyanobutyl vinyl ether)

properties investigated. The simplest phenoxy compound (perfluoro-phenyl vinyl ether) has been reported by Wall and Plummer to polymerize with great difficulty (6). In general, ionically bonded cross-linking agents and groupings that tend to interchange easily (such as ester groups) give good chemical-resistant polymers, but do not give vulcanizates with attractive high-temperature properties because these rubbers suffer high relaxation under stress at elevated temperatures.

Among the more attractive vinyl ether candidate monomers are perfluoro(4-carbomethoxybutyl vinyl ether), perfluoro(4-cyanobutyl vinyl ether), perfluoro(2-phenoxypropyl vinyl ether), and perfluoro(3-phenoxypropyl vinyl ether). Compositions containing these monomers are discussed in this report.

Polymerization is carried out expeditiously in aqueous media using either redox or thermally generated free-radical systems; anhydrous techniques may be also used. In general, there appears to be no gross difference in reactivity rates with different substituents on the vinyl ethers. However, where slight differences were noted, it is likely that differences in solubility in the micelle affect concentration at the polymerization site and thus modify the final content of the third monomer in the polymer. Different crosslinking reactions are used with various substituted vinyl ether structures.   These reactions are discussed under the appropriate copolymer type.

Studies with Perfluoro(4-Cyanobutyl Vinyl Ether)

The synthesis of this monomer involves nine steps starting with perfluoroglutaryl chloride, as indicated in Figure 2 (3).   Most of the steps give good yields except for the selective pyrolysis step (Figure 2, compounds V-VI), where the yields are only 10–20%.   By-products are perfluoro(butenyl vinyl ether) and the hydrogen fluoride adducts of both the desired acid and perfluoro(butenyl vinyl ether).

Polymerization.   Copolymers of tetrafluoroethylene/perfluoro(methyl vinyl ether) and the nitrile (1–4 mole %) have been prepared batch-wise in a stirred autoclave using an aqueous ammonium persulfate or ammonium persulfate-sodium sulfite redox couple system at 40°-100° C. The TFE/PMVE gas mixture was pressured, as required, to maintain the pressure and the nitrile pumped in solution in trichlorotrifluoro-ethane.   After completion of the reaction, polymer was isolated from the latex (25–30% solids) by coagulation using ethanol and aqueous magnesium chloride solution.   It was washed with alcohol/water solutions and dried at 70°C in an oven under nitrogen.   Mass balance indicated that most of the nitrile had been incorporated.

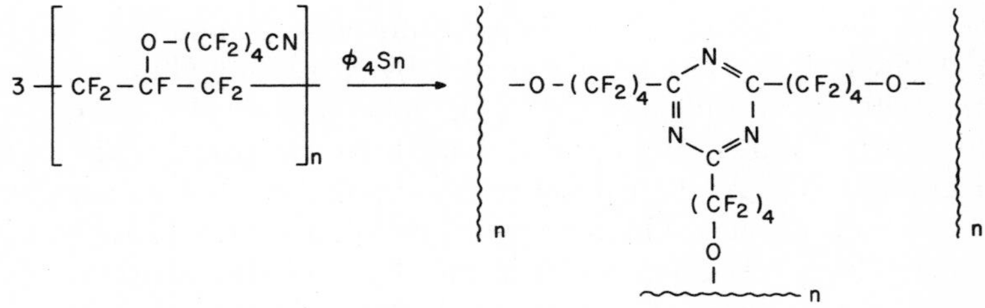

Figure 3.   Crosslinking of polymer through the nitrile group

Crosslinking.   Crosslinking is brought about by the catalytic inter-action of the pendant nitrile groups, perhaps to the triazine, using tetraphenyltin or silver oxide (Figure 3).   The exact form of the cross-link has not been established.   Henne (4) showed monomeric perfluoro carbonitriles form triazine rings in the presence of tetraphenyltin or other catalysts (5).   These triazine structures so formed absorb very strongly at 6.40–6.45μ in the infrared.   Crosslinked gum stocks pre-pared from the perfluoroelastomer (polymer and catalyst) have been examined in the infrared, and other unassigned bands were found, but the characteristic absorption of the triazine structure has not been ob-

RAW TERPOLYMER   100 pts.

MED. THERMAL BLACK   10 pts.

TETRAPHENYL TIN   4.5 pts.

Cure Conditions: 18 hours
160 °C (320°F)
~1000 psi

*Figure   4.   Crosslinking   formulation
for nitrile-containing polymer*

served.   When the viscous nature of the polymer stock is considered, coupled with the fact that the nitrile is present in the polymer to the extent of only 2–4 mole % or less, it appears unlikely that complete conversion to a triazine structure is involved.

In practice, the cure ingredients (*see* Figure 4) are milled into the polymer on a rubber mill; curing is done in a press at 160° C and 1000 psi for 18 hours.   No postcuring is required.

Studies with Perfluoro(4-Carbomethoxybutyl Vinyl Ether)

Several types of crosslinked structures were prepared by copolymerizing vinyl ethers containing 4-carbomethoxyperfluoroalkyl moieties in the structure and using various crosslinking agents to bring about vulcanization.

Synthesis of Perfluoro(4-Carbomethoxybutyl Vinyl Ether).   The parent ester was prepared from perfluoroglutaryl fluoride and hexafluoropropylene oxide by the reactions shown in Figure 2.   This ester had a boiling point of 146°C, and showed characteristic vinyl ether infrared absorptions at 5.42μ, 5.55μ (carbonyl group), and 10.0μ ($-OCH_3$ group).

Acutally, better yields of the monocarboxylic acid esters are obtained by pyrolyzing the half-neutralized acid instead of the dipotassium salt.   Even so, the yield of ester is only about 25%.

Polymerization.   Polymerization is carried out in the same reactor as indicated previously in the standard persulfate-sulfite system.   The presence of the ester (1 mole % charged) in the polymer is indicated by the strong infrared absorption band at 5.55μ, and mass balance indicates that most of the ester is incorporated.

Crosslinking   the   Ester-Containing   Polymer.   Crosslinking   was brought about easily using a conventional fluoroelastomer hexamethylenediamine carbamate cure, or by using *p*-phenylenediamine.   In the samples tested, vulcanizate properties (hardness, elongation) and re-

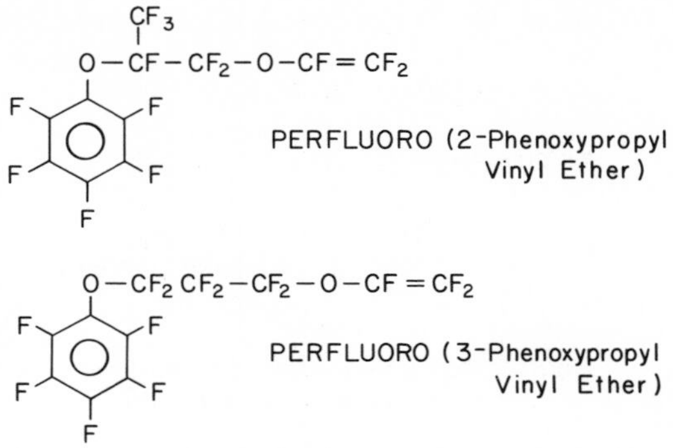

Figure 5.   Monomers containing the perfluorophenoxy group

sistance to chemicals and fluids indicated adequate curing.   Nevertheless, compression set values (70 hours at 121° C) on stocks varied
between 87 and 100%, indicating the extreme mobility of the crosslinking
bond at moderately high temperatures.   Metathetical reactions of the
ester groups, with amines, ammonia, and hydrazine gave model crosslink structures.   Heat-loss studies on these, along with model compounds prepared from the free acid, indicated that carbonyl functions
are not thermally stable enough.

Studies with Monomers Containing the Perfluorophenoxy Group

Two monomers containing the perfluorophenoxy group have been
synthesized and copolymerized with TFE and PMVE.   Structures· of
these monomers are shown in Figure 5.   Polymerization and crosslinking reactions are similar for polymers containing these two isomeric
monomers and, accordingly, these polymers are discussed together.

Synthesis of the Phenoxy-Containing Monomers.   Perfluoro(2-phe-
noxpropyl vinyl ether) is prepared by the route shown in Figure 6 (3).
•The 2-phenoxy compound requires the least-involved synthesis of the
various monomers.   The first two steps were carried out in ordinary
glass equipment.   The last reaction was carried out using a bed of
dried sodium carbonate.

Perfluoro(3-phenoxypropyl vinyl ether) (7) is synthesized by the
six-step route shown in Figure 7.   The first three steps may be carried
out without isolating the products.   The acid chloride product of step
3 is separated by fractional distillation, as is the acid fluoride product of
step 4.   In step 5, the product is separated as the lower of a two-layer
product and is purified by distillation.   The upper layer, a complex of

CsF, the starting acid fluoride, and tetraglyme may be reused as catalyst. Better product yields are obtained with reused catalyst than with virgin catalyst. The last reaction (step 6) is best carried out using a fluidized bed of dried sodium carbonate at 270°-300° C.

Polymerization. Terpolymers of PMVE and TFE with either of the monomers containing a phenoxy group have been prepared in a pressure vessel using an aqueous redox polymerization system. The compositional molar TFE/PMVE ratio in the preferred polymer is about 60/40. The third monomer polymerizes at about the same rate as the PMVE and is fed either neat (as a liquid) or in Freon F-113 solution. Infrared analysis of the band at 10.0μ indicates 75-85% incorporation of the phenoxy compound over the 1-4 mole % monomer change range. One to 2 mole % of the crosslink monomer must be incorporated in the elastomer to ensure good vulcanizate properties.

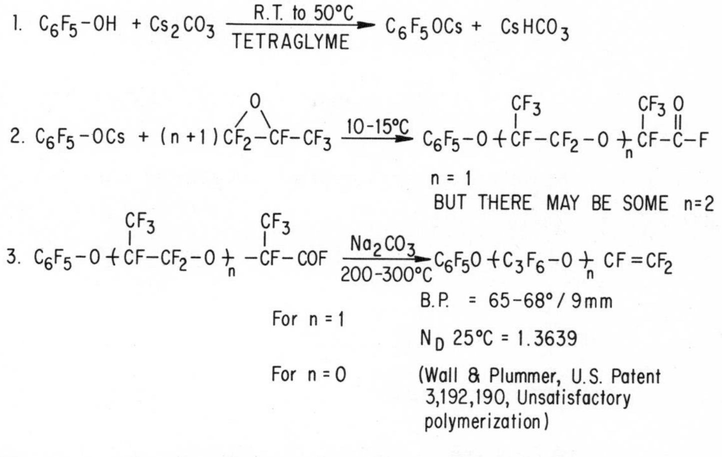

1. $C_6F_5-OH + Cs_2CO_3 \xrightarrow[\text{TETRAGLYME}]{\text{R.T. to } 50°C} C_6F_5OCs + CsHCO_3$

2. $C_6F_5-OCs + (n+1)CF_2\!-\!\overset{O}{\overbrace{}}\!CF-CF_3 \xrightarrow{10-15°C} C_6F_5-O\!\left(\!\overset{CF_3}{\underset{|}{CF}}-CF_2-O\!\right)_{\!n}\!\overset{CF_3}{\underset{|}{CF}}-\overset{O}{\overset{||}{C}}-F$

n = 1
BUT THERE MAY BE SOME n=2

3. $C_6F_5-O\!\left(\!\overset{CF_3}{\underset{|}{CF}}-CF_2-O\!\right)_{\!n}\!\overset{CF_3}{\underset{|}{CF}}-COF \xrightarrow[200-300°C]{Na_2CO_3} C_6F_5O\!\left(C_3F_6-O\right)_{\!n}CF=CF_2$

For n = 1   B.P. = 65-68° / 9mm
              N_D 25°C = 1.3639

For n = 0   (Wall & Plummer, U.S. Patent 3,192,190, Unsatisfactory polymerization)

Figure 6. Synthesis of perfluoro(2-phenoxypropyl vinyl ether)

Crosslinking Perfluorophenoxy-Containing Elastomers. Vulcanization of the perfluoroelastomer containing a phenoxy group may be carried out several ways. One method uses diamines such as p-phenylenediamine, tetraethylenepentamine, or hexamethylenediamine carbamate (Du Pont DIAK No. 1) (3). A typical recipe is shown in Figure 8.

The rubber and the compounding ingredients are mixed on a conventional rubber mill, and the desired shape molded in a press at 500-1000 psi on the shaped piece for 30 minutes at 160° C. The piece is then removed from the press and postcured stepwise over five to six days to complete vulvanization. Our present concept of this crosslinking reaction involves reaction of the diamine (Figure 9) with the active fluorine atom para to the oxygen linkage on the aromatic ring to liberate hydrogen fluoride (11). This, in turn, reacts with magnesium oxide to

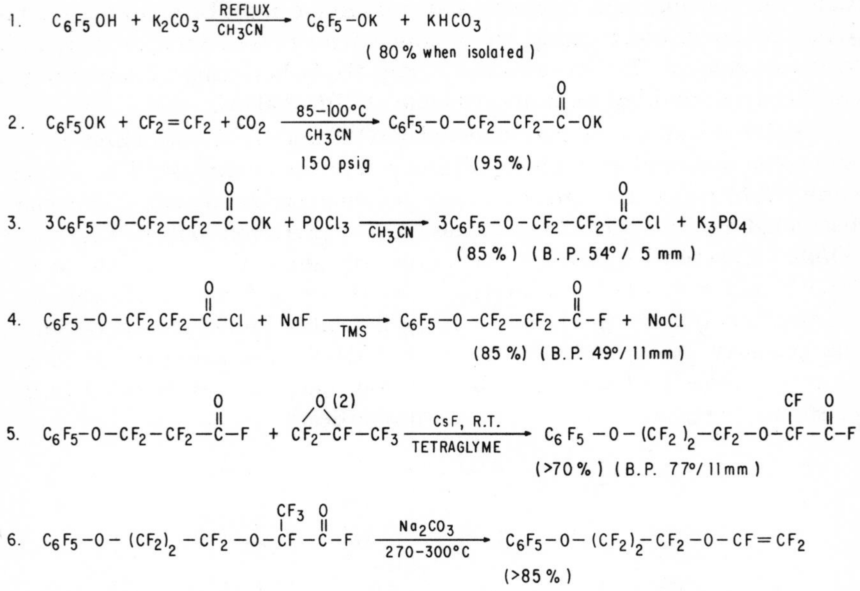

1. $C_6F_5OH + K_2CO_3 \xrightarrow[CH_3CN]{REFLUX} C_6F_5-OK + KHCO_3$

(80% when isolated)

2. $C_6F_5OK + CF_2=CF_2 + CO_2 \xrightarrow[CH_3CN]{85-100^\circ C} C_6F_5-O-CF_2-CF_2-\overset{\overset{\displaystyle O}{\|}}{C}-OK$

150 psig   (95%)

3. $3C_6F_5-O-CF_2-CF_2-\overset{\overset{\displaystyle O}{\|}}{C}-OK + POCl_3 \xrightarrow{CH_3CN} 3C_6F_5-O-CF_2-CF_2-\overset{\overset{\displaystyle O}{\|}}{C}-Cl + K_3PO_4$

(85%) (B.P. 54°/ 5 mm)

4. $C_6F_5-O-CF_2CF_2-\overset{\overset{\displaystyle O}{\|}}{C}-Cl + NaF \xrightarrow{TMS} C_6F_5-O-CF_2-CF_2-\overset{\overset{\displaystyle O}{\|}}{C}-F + NaCl$

(85%) (B.P. 49°/11mm)

5. $C_6F_5-O-CF_2-CF_2-\overset{\overset{\displaystyle O}{\|}}{C}-F + CF_2-\overset{\displaystyle \overset{O\,(2)}{\diagdown}}{CF}-CF_3 \xrightarrow[TETRAGLYME]{CsF,\ R.T.} C_6F_5-O-(CF_2)_2-CF_2-O-\overset{\overset{\displaystyle CF}{|}}{CF}-\overset{\overset{\displaystyle O}{\|}}{C}-F$

(>70%) (B.P. 77°/11mm)

6. $C_6F_5-O-(CF_2)_2-CF_2-O-\overset{\overset{\displaystyle CF_3}{|}}{CF}-\overset{\overset{\displaystyle O}{\|}}{C}-F \xrightarrow[270-300^\circ C]{Na_2CO_3} C_6F_5-O-(CF_2)_2-CF_2-O-CF=CF_2$

(>85%)

Figure 7.   Synthesis of perfluoro(3-phenoxypropyl vinyl ether)

	PARTS by WEIGHT
POLYMER	100
MEDIUM THERMAL BLACK	20.0
MAGNESIUM OXIDE	15.0
HEXAMETHYLENE DIAMINE CARBAMATE	1.5

Figure 8.   Diamine formulation for curing perfluorophenoxy-containing polymer

form water and magnesium fluoride. Under the conditions required to complete the vulcanization, water must be expelled from the molded piece, and, because of the low permeability of the rubber to moisture, the step postcure must be carried out slowly to prevent sponging.

Alternately, aromatic bisnucleophiles may be used to cure these perfluorophenoxy-containing polymers. Examples of typical crosslinking agents are shown in Figure 10. Potassium salts are shown, but calcium salts are also effective. These crosslinking agents require an accelerator to function well. In general, the accelerators found to be most efficient are polyethers (Figure 11) such as polyethylene ether glycols— $CH_3-O-(CH_2-CH_2)_n-OH$—of molecular weight about 350 (Carbowax

350) (8), or the cyclic polyether compounds of "crown" structure (9). A typical formulation is shown in Figure 12.

The role of the accelerators is not known, but they may serve as solubilizing agents, complexing agents, or ionization agents. The complete curing reaction is quite slow, and the content of crosslinking sites is low so that long post cures—up to five days—are required. The last 24 hours of the cure is carried out at 290° C under nitrogen to obtain the best high-temperature properties.

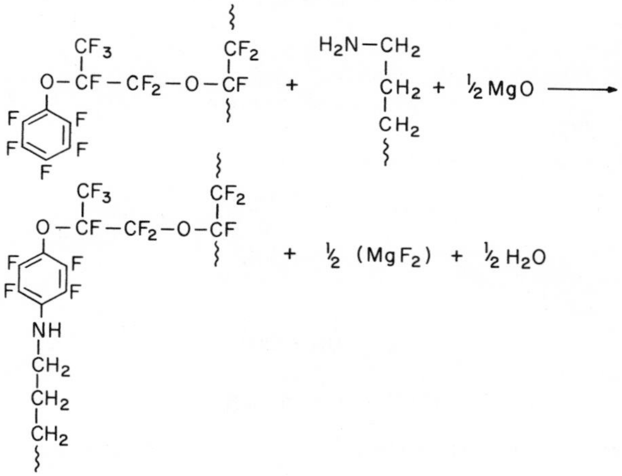

Figure 9. Diamine curing of phenoxy-containing polymer

Raw Polymer Properties

The physical properties of raw terpolymer are essentially identical with that of the dipolymer described by Barney *et al.* (1), with a few exceptions. Low-temperature characteristics are essentially unchanged, but uncured polymers show some loss in thermal stability attributable to the third monomer. For example, a perfluorophenoxy-containing polymer lost about 12% of its weight after heat aging in an air circulating oven at 316° C in six days, whereas the dipolymer control lost only 6.5% under similar conditions.

Because of the similarity of the third monomers investigated, raw polymer density is essentially constant from polymer to polymer: 2.04 ± 0.01. Cured and postcured gum stock has a density of 2.01 and black stock containing 10 phr. SAF black has a density of 2.02.

Raw polymer is insoluble in most fluids, making the determination of solution viscosity very difficult. Solutions of 0.2 g/100 g solvent may be prepared in 2,2,3-trichloroheptafluorobutane (Halocarbon 437, by

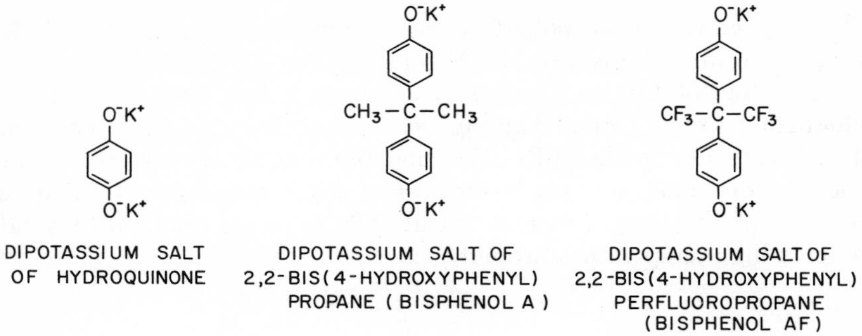

DIPOTASSIUM SALT
OF HYDROQUINONE

DIPOTASSIUM SALT OF
2,2-BIS(4-HYDROXYPHENYL)
PROPANE (BISPHENOL A)

DIPOTASSIUM SALT OF
2,2-BIS(4-HYDROXYPHENYL)
PERFLUOROPROPANE
(BISPHENOL AF)

Figure 10. Typical bisnucleophiles used in curing phenoxy-containing perfluoroelastomer

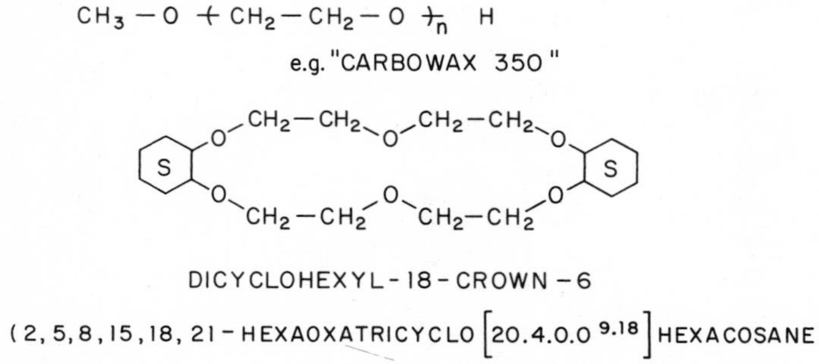

$$CH_3 - O + CH_2 - CH_2 - O \;)_n \; H$$

e.g. "CARBOWAX 350"

DICYCLOHEXYL-18-CROWN-6

$(2,5,8,15,18,21-HEXAOXATRICYCLO [20.4.0.0^{9.18}] HEXACOSANE)$

Figure 11. Accelerators for perfluoroelastomer curing

Halocarbon Products, Inc.) or in a few other chlorofluorocarbons containing 2-3% by weight of a polar cosolvent such as methanol, trichloroacetic acid, or glyme. In these solvents, determinations of solution viscosity indicate that inherent viscosities (η_{inh}) of 0.4 or greater are necessary to obtain adequate vulcanizate properties.

Properties of the Cured Polymer Stocks

Experimental studies to date indicate that there is little basic difference in the resistance to chemicals or swelling by fluids between compositions containing the different crosslink monomers. This is probably because the cure site monomer, and even the crosslinking agent in the polymer, represent only a small part of the finished vulcanizate. Furthermore, the inherent resistance of the polymer to various fluids and to chemical attack, coupled with its low permeability, contribute to the similarity of materials cured with various agents. Only the most drastic tests show differences in performance.

A brief tabulation of the vulcanizate properties to indicate the level of performance that can be achieved with these new perfluoroelastomers is in order (*10*):

Figure 13 tabulates some of the physical properties. The data shown are for black stocks.

Figure 14, illustrates the outstanding resistance of black vulcanizates to heat aging in air (thermal oxidation).

	BLACK STOCK	GUM STOCK
POLYMER	100	100
SUPER ABRASION FURNACE BLACK	10	–
MAGNESIUM OXIDE	4	–
ACCELERATOR	3	3
BISPHENOL	3	$4\frac{1}{2}$

Cure : 30 min at 177 °C

Oven Cure : ROOM TEMP to 290 °C – 5 DAY STEP – N_2

Figure 12. Typical cure formulation for curing phenoxy-containing perfluoroelastomer with bisnucleophiles

90 DUROMETER STOCK   20 phr SAF BLACK + CURATIVES

Cure 30 mins. / 350 °F + POSTCURE TO 400 °F (500 °F for COMP. SET)

ORIGINAL PHYSICAL PROPERTIES

STRESS / STRAIN at 75 °F	
M_{100} , psi.	1350
T_B , psi.	2720
E_B , %	160
HARDNESS (Durometer A)	89

COMPRESSION SET – METHOD B

70 hrs. at 121 °C (250 °F)	23%
70 hrs. at 288 °C (550 °F)	45%

CLASH – BERG STIFFNESS TEMP.
(10,000 psi TORSIONAL MODULUS)   +28 °F

BRITTLE TEMPERATURE   –38 °F

TEMPERATURE RETRACTION
T-10	+30 °F
T-50	46 °F

TEAR D-470 (pli)   13

Figure 13. Typical physical properties of perfluoroelastomer black vulcanizates

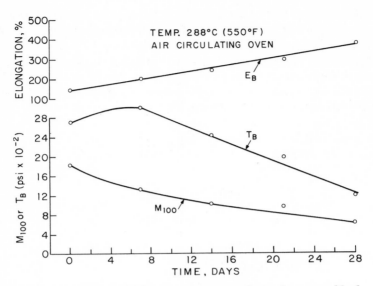

Figure 14.   Effect of heat aging on perfluoroelastomer black
(10 phr SAF) vulcanizate tensile properties

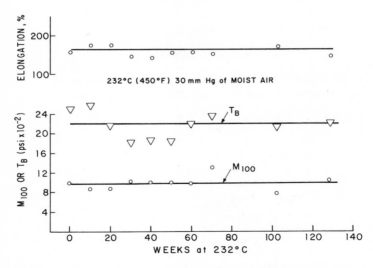

Figure 15.   Effect of heat aging under conditions of operation
for high-performance aircraft

Figure 15 shows heat aging characteristics at 30 mm air pressure
and 232° C—conditions similar to what a high-performance aircraft
might encounter in flight.

Figure 16 illustrates the excellent resistance of black vulcanized
stocks to swelling with various fluids.

In addition to black, the elastomer may be compounded with other common fillers such as silica, clay, and asbestos. However, in general, their reinforcing properties are inferior to carbon. For certain uses, it is advantageous to compound the rubber without fillers. This gum vulcanizate has poorer physical properties than loaded stock but is better in chemical resistance. Both reinforced and gum recipes may include perfluoro greases or oils as plasticizers to soften vulcanizates without affecting appreciably chemical resistance.

Figure 17 tabulates the electrical properties of the gum stock in comparison with those of Teflon fluorocarbon resin.

	WT. INCREASE,%	TENSILE RETD.,%	ELONG. RETD.,%
JP-5 JET-FUEL	0	100	100
BENZENE	1.3	80	100
CARBON TETRACHLORIDE	3.2	75	100
2B ALCOHOL	0	100	100
ACETONE	1.0	73	100
ETHYL ACETATE	1.2	73	100
PYRIDINE	1.0	96	100
ACETIC ANHYDRIDE	0.5	94	100
SODIUM HYDROXIDE (46%)	0	100	100

Figure 16. Resistance of black vulcanizate perfluoroelastomer (10 phr SAF black) to various fluids at 25°C and seven-day immersion

	ELASTOMER [1]	TFE [2]
DIELECTRIC CONSTANT "E"	2.8–3.2	2.1
DISSIPATION FACTOR "D" (50% RH, 73°F)	1×10^{-3}	3×10^{-4}
D.C. RESISTIVITY "ρ"	10^{18} ohm-cm	10^{18}
DIELECTRIC STRENGTH (BREAKDOWN)	> 2000 volts/mil	400

[1] GUM STOCK -VULCANIZED

[2] PLASTICS WORLD (1965).

Figure 17. Dielectric properties of gum perfluoroelastomer

Safety Notes

(1) Many of perfluoro compounds mentioned are toxic and should be handled only by competent investigators in well ventilated areas.

(2) The polymerization work reported herein requires compression of tetrafluoroethylene-perfluoro(methyl vinyl ether)mixtures. This operation must be considered potentially hazardous and should be carried out only in adequately barricaded areas.

Acknowledgments

The authors thank A. F. Breazeale, D. F. Brizzolara, D. B. Pattison, and H. J. Stinger for their contributions to this study.

Literature Cited

1. Barney, A. L., Keller, W. J., van Gulick, N. M., *J. Polym. Sci.,* A-1 (1970) **8,** 1091.
2. Gladding, E. K., Sullivan, R; U.S. Patent **3,546,186** (1970); Anderson, D. G., Gladding, E. K., Sullivan, R.; French Patent **1,527,816** (1968).
3. Pattison, D. B.; U.S. Patent **3,467,638** (1969).
4. Henne, A., Pelley, R. L., *J. Amer. Chem. Soc.* (1952) **74,** 1426.
5. Graham, T. L., *Rubber Age* (1960) 43; Fritz, C. G.; U.S. Patent **3,317,484** (1967).
6. Wall, L. A., Plummer, W. J.; U.S. Patent **3,192,190** (1965).
7. Quarles, R. W., Jr., French Patent **71.09113** (1971); Brizzolara, D. F., Quarles, R. W., Jr., French Patent **71.09114** (1971).
8. Barney, A. L., Honsberg, W.; U.S. Patent **3,524,836** (1970).
9. Barney, A. L., Honsberg, W.; U.S. Patent **3,580,889** (1971).
10. Barney, A. L., Kalb, G. H., Khan, A. A., *Rubber Chem. and Tech.* (1971) **44,** 660.
11. Brizzolara, D. F.; this laboratory; unpublished results based on model compound studies indicate primarily para position attack.

RECEIVED April 13, 1972.

Novel Heat Resistant Plastics from Hydrogenation of Styrene Polymers

J. F. PENDLETON and D. F. HOEG

Roy C. Ingersoll Research Center, Borg-Warner Corp.,
Des Plaines, Ill. 60018

E. P. GOLDBERG

Xerox Corp., Rochester Research Center, Webster, N.Y.

The combination of polymer preparation by anionic tech-niques and hydrogenation technology based on soluble transition metal catalysts has allowed the facile hydrogena-tion of polystyrene and styrene–diene block copolymers. Problems encountered in hydrogenation of polymers using conventional techniques have been avoided. Polystyrene is hydrogenated to poly(vinylcyclohexane) (PVCH) which has a heat distortion temperature 60°C higher than polystyrene. PVCH, however, has low impact strength. By hydrogenat-ing styrene–diene block copolymers, materials with heat and impact resistance resulted. By varying the level of polystyrene in the diene block copolymers products of hydrogenation ranged from rubbery to rigid materials all of which were clear, tough, and heat resistant.

Chemical modification of polymers (1), such as halogenation, epoxida-tion, and chlorosulfonation, has given the industry useful products which in most cases could not be made directly from monomer poly-merization. Hydrogenation, a more esoteric form of polymer modifica-tion, while being as old as polymer chemistry, has not given commer-cially useful polymers until recently. Staudinger, in the 1920's, used the hydrogenation of rubber (2) and low molecular weight polystyrene (3) to prove that the macromolecules were composed of long chains with primary valence bonds as opposed to the current theories at that time which considered "polymers" as association complexes of low molec-ular weight species (4). Since that time hydrogenations of polystyrenes and polydienes have been studied in the hope of obtaining commercially attractive polymer derivatives (5, 6, 7, 8, 9). Two reviews in this area by Moberly (10) and Wicklatz (11) discuss the development of new techniques both in synthesis and catalysis for the hydrogenation of poly-dienes and styrene–diene copolymers. While the bulk of research discussed in these reviews has been involved with polydienes, some interesting work with non-diene polymers should be mentioned. Polyacrylonitrile has been hydrogenated by J. H. Parker (12) to the polyallylamine using a 15% nickel catalyst in the presence of ammonia. Gluesenkamp (13) reported the partial hydrogenation of poly(vinyl chloride) in dimethyl acetamide again with a high level of catalyst (Pd/C). Steinhofer (14) and Warner (15), using high concentrations of nickel catalysts, hydrogenated polystyrene homopolymer to poly(vinylcyclohexane). As observed in all the references, the use of noble metal and typical transition metal hydrogenation catalysts required long reaction times, high catalyst concentrations, and high temperature. This was accompanied by problems in catalyst removal and in most cases some molecular breakdown of the polymer during hydrogenation. Problems are compounded when the polymer to be hydrogenated is crosslinked or not completely soluble in the hydrogenation solvent. The recent use of so-called soluble transition metal catalysts made by the reaction of alkylaluminums and transition metal salts is described in Moberly's review, and these systems, at least in polydiene and styrene diene copolymers, avoid the problems encountered with typical heterogeneous catalysts. Work in our laboratory confirms that residence times are short, catalyst concentrations are low, and product workup is easy. In fact, the activity of such systems is truly catalytic in that as much as 4 kg of a styrene–butadiene copolymer can be hydrogenated completely using a soluble catalyst containing only 1 gram of cobalt.

The impetus for this research was generated from the extraordinary properties established for hydrogenated amorphorus polystyrene—*i.e.*, poly(vinylcyclohexane) (PVCH). These properties are shown in Table I. By complete hydrogenation of polystyrene, density is decreased by 10%, hardness is significantly improved. clarity is maintained, and with this heat resistance is dramatically improved. An increase of 60°C in ASTM heat distortion is demonstrated; this, of course, is reflected in the high tensile strengths at elevated temperatures shown in Table I. Unfortunately, the brittle character of polystyrene system was not improved on hydrogenation. This was the starting point for this research—to develop a tough product while maintaining the inherent heat resistance of the PVCH.

Since vinylcyclohexane could not be polymerized by conventional free radical initiation, in our laboratory, the option of polymerization in

Table I. Physical Properties of Polystyrene (PS) vs. Poly(vinylcyclohexane) (PVCH)

Property	PS	PVCH
Specific gravity	1.04	0.94
Rockwell hardness, "R" scale	116.0	125.5
$\eta_{sp/c}$, dl/gram	0.78 (C_6H_6)	0.61 (C_6H_{12})
Molecular weight ($\overline{M}_w$) (light scattering)	265,000	235,000
Dielectric constant	2.58	2.56
Ultimate tensile $\times 10^3$ psi		
25°C	4.5	3.9
60°C	2.9	1.9
80°C	2.0	1.5
100°C	0.05	1.4
120°C	—	1.2
Heat distortion temperature, °C (ASTM-648, 264 psi)	81.5	142.5
Dynstat impact (½" unnotched)	2.9 kg cm/cm²	2.9 kg cm/cm²

the presence of rubber as done to prepare impact styrene was not viable. Melt blending of PVCH with rubbery polymers did not impart the degree of toughness that could be obtained in melt blending polystyrene with rubbers. The fact that no impact modification was obtained even when the blended rubber was almost completely saturated indicated that no grafting on the rubber was accomplished as is assumed in polystyrene–rubber blending to produce impact polystyrene. These factors plus the lack of an economical source of vinylcyclohexane showed that the best method of modifying PVCH was to modify polystyrene and hydrogenate this to the PVCH derivative.

We believe that a rubber-modified polystyrene could not be hydrogenated efficiently. Even a styrene–butadiene emulsion copolymer with low gel content could only be partially hydrogenated and then only very slowly followed by difficult catalyst removal.

Soluble copolymers of styrene were considered as candidates for hydrogenation to impact PVCH materials. Anionic diene copolymers with styrene were chosen for study because the structure of polydiene portions could be controlled to give flexible rubbery segments on hydrogenation.   The polydiene structures listed in Table II all give flexible to rubbery products on hydrogenation.

Polyisoprene on hydrogenation gives a rubber directly—an alternating ethylene–propylene polymer.   Polybutadiene can give polyethylene on hydrogenation if it is all 1,4 in structure or a variety of flexible-to-rubbery ethylene–butene copolymers as the 1,2 content of the polybutadiene is increased.   These polydiene structures can be incorporated as segments in anionic styrene copolymers.

Table II.   Hydrogenated Diene Polymers

$$
\begin{array}{c}
CH_3 \\
| \\
-CH_2C{=}CHCH_2-_n
\end{array}
\xrightarrow{H}
\begin{array}{c}
CH_3 \\
| \\
-CH_2CH \ CH_2CH_2-_n
\end{array}
$$

Poly(1,4-isoprene)   Alternating EP rubber

$$
-CH_2CH{=}CHCH_2-_n \xrightarrow{H} -CH_2CH_2CH_2CH_2-
$$

Poly(1,4-butadiene)   Polyethylene

$$
-CH_2CH{=}CH-CH_2-_n \ -CH_2CH-_m \xrightarrow{H}
$$
$$
\begin{array}{c}
| \\
CH \\
\| \\
CH_2
\end{array}
$$

Random 1,4-1,2-polybutadiene

$$
-CH_2CH_2CH_2CH_2-_n \ -CH_2CH-_m
$$
$$
\begin{array}{c}
| \\
CH_2 \\
| \\
CH_3
\end{array}
$$

Random ethylene–butene elastomer

In addition to the variation in diene composition and microstructure, anionic techniques also allow a variety of block copolymer structures. We investigated three basic types of block structure.   Using styrene and butadiene as the model system, two block structures studied were pure di- and triblock prepared by sequential addition of the monomers in a butyllithium initiated system, giving styrene–butadiene (SB) and styrene–butadiene–styrene (SBS) block polymers.   In addition, mixed block polymers (SB^M) were hydrogenated.   This type of starting polymer was prepared by alkyllithium initiation with both monomers present. The first block is a mixture of mostly polybutadiene with some of the styrene randomly distributed in this segment (10 to 35% of total styrene charged, depending on composition) with the remaining styrene only comprising the second block (16, 17).   The letter designations used here for the various products are based on the designation for the un-hydrogenated products—SBS, SB, or SB^M for styrene–butadiene systems where B is the typical 90% 1,4- /10% 1-2-polybutadiene from butyllithium initiation in hydrocarbons. For polybutadiene of higher 1,2 content the designation is, for example, SB (1,2)S. In all high vinyl products the (1,2) percent in the polybutadiene segment is about 50%. The number preceding the letters is the overall styrene content prior to hydrogenation. The hydrogenated product carries an R after the composition identification, and this refers to complete hydrogenation of both the diene and styrene portions. For example as 25SB^M-R is a fully hydrogenated mixed block polymer of 25% styrene and 75% butadiene.

The hydrogenated polymers in all ranges were transparent and had heat stability far superior to the styrene–butadiene counterparts. The inherent thermal oxidative stability was demonstrated by repeated extrusion of unstabilized products without loss of clarity or decrease in viscosity. The level of styrene content studied was from 10% to 90%, and the properties of the hydrogenated derivatives generally ranged from tough rubbery products at low styrene to tough, rigid, heat resistant products at high styrene content. The products obtained in these two areas are described below.

Laboratory Procedure

All operations were carried out in the absence of air under argon.

Styrene Block Polymer Synthesis. GLASSWARE. All glassware was dried at 130°C in a forced air oven overnight and cooled to room temperature under an argon purge. The 12 oz and 28 oz "pop" bottles were capped immediately after cooling under an argon purge, with a neoprene liner and a metal perforated crown cap. The 12-oz bottles are borosilicate glass, and the 28-oz bottles were soft glass. In all cases, contents were added to the bottles by syringe needle.

SOLVENTS. *Tetrahydrofuran (THF)*. THF (Dupont) was distilled in a simple distillation setup, which had been flamed out under argon. To 1500 ml of the THF, which was passed through molecular sieves, we added 20 ml. styrene and enough 1.5N butyllithium solution in hexane to give a permanent yellow collor. A small forerun was discarded, and the THF distilled directly into 28-oz bottles, which were capped and pressured with argon.

Cyclohexane. Pure grade cyclohexane (Phillips Petroleum Co.) was passed through activated Linde 3A molecular sieves under argon into a stainless steel oxygen bomb (9 gal). It was dispensed under pressure through a valve connected to polyethylene tubing with a syringe needle attached.

MONOMERS. *Styrene*. Styrene (Eastman white label), 1200 ml, was placed in a dry 2-liter R.B. flask containing a Teflon-covered magnetic stirrer. While stirring under argon, diethyl ether, 200 ml (anhydrous), and benzophenone, 12 grams, were added and then 1.0 gram of sodium in small pieces. On stirring overnight, a deep blue color develops, indicating that all impurities are purged. Without stopping the stirring, the flask was attached to a dry, vacuum distillation setup. The system was then evacuated and filled with argon twice, and the vacuum was brought eventually to 40 mm as the vapor pressure of the ether permits. The ether was collected in the dry ice trap and a forerun of 25-50 ml was collected at 50°C and 40 mm. Pure styrene was then collected (50°C/40 mm), yielding approximately 1 liter. This was transferred to 12-oz capped bottles by syringe needles, purged with argon, and stored in a dry ice chest.

Butadiene. Phillips special purity butadiene (99.5 mole %) was distilled from the cylinder into a dry ice chilled 1700-cc bomb. This bomb was then inverted and connected to a 4' x 2" stainless steel column packed with 3A molecular sieves. (4A sieves cause polymerization; 3A do not.) The liquid butadiene was passed through this column and metered out slowly (150 grams in 15 minutes) through a vernier needle valve connected to a syringe needle into the polymerization bottles.

Isoprene. Matheson pure grade isoprene was passed slowly through 3A molecular sieves into a dry R.B. flask. To this (500 ml) we added 50 ml of a 25% triisobutylaluminum solution in hexane. The isoprene was distilled in a simple distillation setup and collected in 12-oz bottles which are capped and maintained at 10 psi argon. Addition to polymerization bottles was accomplished by syringe.

PURGING SOLUTION AND CATALYST PREPARATION. To a dry, 12-oz capped bottle we added 100 ml of benzene, 5 ml of α-methylstyrene and 5 ml of 1.5N butyllithium in hexane. This mixture was warmed to 50°C for one-half hour or until the deep red α-methylstyrene anion color formed. It was kept under 10 psi argon and stored with the styrene in the dry ice chest.

The catalyst solution was prepared by adding 25 ml 1.5N butyllithium (Foote Mineral Co., in hexane) to 250 ml of argon-degassed cyclohexane in a 12-oz bottle. The active butyllithium concentration was determined by the method of Gilman (18), except that THF was used instead of diethyl ether.

The reaction flasks (50 ml Erlenmeyer) were dry, argon-filled, and capped with serum stoppers. To each was added 10 ml THF and 1 ml benzyl chloride. Four ml of catalyst solution were added to two of these and 8 ml to two more. After at least one minute, the solutions were poured into distilled water (100 ml), and the LiOH was titrated with standard acid to a phenolphthalein endpoint. The difference between the 4 ml and 8 ml aliquot samples is the inactive content of the catalyst solution. This takes into account any impurities which may have been present in the THF or benzyl chloride. Total lithium content was determined by quenching 4 ml of the catalyst solution in 100 ml of water and titrating with standard acid. The molarity of the catalyst solution in active butyllithium is simply the total LiOH acid titre less the inactive LiOH titre.

POLYMERIZATION PROCEDURE. The bottles used to store styrene, isoprene, purging solution, catalyst solution, cyclohexane, and THF are all maintained under approximately 15 psi argon pressure. In removing samples by syringe, the bottles are inverted, and the solutions are pressured into the syringe. This eliminates any chance of contamination by air seepage around the barrel of the syringe, as sometimes happens in normal syringe operation.

All laboratory polymerizations were run in 28-oz bottles containing a Teflon-covered magnetic stirring bar. Cyclohexane, 550 ml, was pressured into the bottle, and the solvent was degassed by inserting a long, 18 gage needle to the bottom of the bottle and blowing argon through the rapidly stirring solvent and out an 18 gage needle at the top. At pressures between 10 and 25 psi, the maximum flow rate in such a system is approximately 3 liters/minute. This flow continued for at least 15 minutes.

The proper amount of styrene was added by syringe, and argon purging was continued for a few minutees longer. The mixture was then cooled to approximately 10°C, and, while stirring, the purging solution was added carefully until a slight color change was observed—water-white to a light yellow or tan. The bottle was then shaken to ensure that all impurities were purged.

If this is to be a mixed block synthesis, butadiene is added at this point and then the catalyst. The following procedure describes a pure ABA block synthesis.

With the styrene purged, catalyst is added by syringe in an amount which depends on the final molecular weight desired according to the following equation:

$$\text{Molecular weight} = \frac{\text{grams monomer}}{\text{moles catalyst}}$$

The bottle is then placed in a 50°C water bath and is stirred by a magnetic stirring motor under the bath for 3 hours. It is removed and cooled to near room temperature, weighed, and butadiene is added from the butadiene drying column. When the desired weight of butadiene has been introduced, the bottle is returned to the bath, and the polymerization is allowed to proceed for 5 hours.

For the last block, a styrene-in-cyclohexane solution is prepared and purged similar to the starting solution, and the proper amount is added by syringe directly into the polymerization bottle at 50°C. This portion of the polymerization is allowed to proceed for an additional 3 hours and is terminated by the addition of a few milliliters of methanol. The polymers are precipitated in methanol in a Waring Blendor, stabilized with 1% N-phenyl-β-naphthylamine and dried in a vacuum oven. If the product is hydrogenated, the quench with methanol is not done and the cyclohexane solution of the polymer is used directly.

Hydrogenation Catalyst Preparation. A 0.081M solution of hydrogenation catalyst (based on cobalt) was prepared by adding 23.6 grams of cobalt(II) 2-ethylhexanoate cyclohexane solution (12.0% cobalt w/w, Harshaw Chemical Co.) over a period of 90 minutes to a solution of 18.8 grams (0.165 mole) of triethylaluminum in 495 grams of cyclohexane (Texas Alkyls). Addition is to a capped bottle with venting. The aluminum-to-cobalt ratio was 3:1. A nickel-based system can be prepared by substituting nickel(II) 2-ethylhexanoate for the cobalt octoate.

Hydrogenation Procedure. Ninety-five pounds of a cyclohexane solution containing 12.5 pounds of a 40 SBS polymer ($\eta_{sp/c} = 0.79$ dl/gram, benzene 23°C) were charged to a 20-gal autoclave. The reactor was pressurized twice to 50 psig with hydrogen and 750 ml of a 0.19M cobalt catalyst was added. The hydrogenation was conducted at a hydrogen pressure of 3500 psi at 250°C for 80 minutes. The product in solution at 50°–60°C was mixed with 10% aqueous nitric acid to extract catalyst residues and precipitated in methanol. The white product had a viscosity of 0.85 dl/gram in Decalin at 135°C ($\eta_{sp/c}$). There was no unsaturation, either aliphatic or aromatic, detected by infrared analysis on a Perkin-Elmer recording grating spectrophotometer. A film was cast from hot cyclohexane onto NaCl plates, and no absorption was observed at 960 cm^{-1} (*trans*-1,4-polybutadiene), 910 cm^{-1} (1,2-polybutadiene), or 690–695 cm^{-1} (polystyrene). Prepared standards showed that less than 0.5% aliphatic or aromatic unsaturation can be readily observed.

Table III.   Retention of Tensile Strength at Elevated Temperatures

| *Polymer* | $\eta_{sp/c}$ | *Tensile* | |
		T, °C	*psi*
1,4-PBde-R	1.73	25	3160
		75	1180
10SB^M-R	1.72	25	3440
		75	1940
15SB^M-R	1.64	25	4240
		75	2740
20SB^M-R	1.74	25	2770
		75	2000
25SBS-R	1.62	25	3030
		75	2420
40SBS-R	1.10	25	4500
		100	1500

Results

Flexible Hydrogenated Styrene–Diene Polymers. These hydrogenated styrene butadiene polymers with less than 50% styrene are tough, puncture resistant, clear, flexible plastics; certain members have true elastomeric properties. The heat resistance imparted by the PVCH segments is shown in Table III. As indicated earlier the designation -R shows complete hydrogenation of all unsaturation, both aromatic and alphatic. The retention of tensile strength at elevated temperatures is directly related to the increase in PVCH content. At 40% PVCH (for the 40 SBS-R) a tensile strength of 1500 psig was observed even at 100°C.

The effects of block structure and microstructure and the overall effect of hydrogenation in the low styrene containing polymers are illustrated in Table IV. The data show that the unhydrogenated SB

Table IV. Effect of Hydrogenation and Block Structure in 25% Styrene–Diene Polymers

No.	Polymer	$\eta_{sp/c}$, dl/gram	Yield, psi	Ultimate, psi	Elongation, %	IER,[a] %
				Tensile		
1	25SB	0.98	—	30	<5	—
2	25SB-R	1.25	1820	3420	600	72
3	25SB[M]	0.84	—	110	<5	—
4	25SB[M]-R	0.99	1300	4380	350	82
5	25SBS	1.32	190	>650	>1300	High
6	25SBS-R	1.62	1780	4550	800	75
7	25SB(1,2)S-R	.94	180	>2570	>1300	94
8	25SIS-R	1.39	240	1390	1000	97

[a] Immediate elastic recovery from 100% extension

Table V. Physical Properties Rigid Hydrogenated SB Polymers

No.	Polymer	$\eta_{sp/c}$, dl/gram	Yield, psi	Ultimate, psi	Elongation, %	Flex. Mod. psi $\times 10^3$	HDT,[a] °C	Impact,[b] ft lbs/in
				Tensile				
1	56SB[M]-R	1.67	2200	3500	445	—	114	NB
2	70SB[M]-R	1.47	3400	3800	100	150	129	1.9
3	75SB[M]-R	2.10	3560	3610	20	160	136	1.9
4	80SB[M]-R	1.13	—	4800	10	260	138	.9
5	90SB[M]-R	.89	—	5560	5	—	132	.2
6	70SB-R	.86	—	4730	5	320	140	.4
7	70SBS-R	1.29	—	4690	6	280	138	1.1
8	70SB(1,2)S-R	1.31	3480	3190	9	200	140	1.3
9	75SBS-R	1.10	—	4750	<5	330	138	1.0
10	80SBS-R	0.88	—	5180	7	365	137	0.6

[a] Micro-Test. ASTM-648 equivalent for 70 SBS-R = 104°C. 1/8 × 1/4 × 1/2″ sample 10 mil deflection at 264 psi as described by Chang and Goldberg (*20*).
[b] Notched Izod impact strength

Table VI. Heat Distortion Temperature *vs.* Loading

	Polymer			
	70SBS-R	*75SB^M-R*	*PP^a*	*P4-MP^b*
Vicat softening temp., °F	293	279	311	383
Heat distortion temp., (ASTM-648) 264 psi, °F	217	202	145	127
66 psi, °F	252	230	—	—

^{a} Polypropylene (Hercules-Profax)
^{b} Poly-4-methylpentene (ICI-TPX)

and SB^M (Nos. 1 and 3) diblock polymers are rubbery but have low breaking strength and would require compounding and vulcanization to be useful elastomeric materials. However, their hydrogenated counterparts (Nos. 2 and 4) have excellent physical properties and as indicated earlier (Table III) have resistance to heat superior to the unhydrogenated polymers. They are plastic in nature having relatively high yield tensiles and fairly low elastic recoveries. The unhydrogenated triblock polymer (No. 5) is truly elastic in the unvalcanized state. This has also been described elsewhere (*19*). The hydrogenation of the triblock polymer (No. 6), however, gives a material similar to the hydrogenated diblock polymers, and thermoplastic elastomer properties are not observed until the microstructure of the central diene segment is changed to high 1,2 in the starting block polymer (No. 7). Rather than a polyethylene-like central segment as in No. 6, No. 7 has a rubbery central segment; in this case being specifically a 50/50 random ethylene–butene copolymer. This effect has also been reported (*8*). Polymer 8 has thermoplastic elastomer properties even though the central block in the starting material is 1,4-polyisoprene because the hydrogenated derivative is an alternating ethylene propylene copolymer.

Rigid Hydrogenated Styrene–Diene Polymers. While the effects of changing block structure in the low styrene unhydrogenated polymers were quite dramatic, this is not the case with the high styrene or rigid copolymers. The effect of hydrogenation was dramatic, however, and is reflected in the heat resistant properties of this series. Table V lists properties of various rigid hydrogenated styrene diene polymers. As expected, increasing the PVCH level increases the heat distortion and flexural moduli in all block structures. At a given PVCH level the SB-R polymer (*cf.* Nos. 2, 6 and 7) has the highest flexural modulus but also has the lowest impact strength. The SBS-R polymers (7 and 10) are stiffer than the SB^M-R polymers (2, 3 and 4) at the same PVCH content, but the SB^M-R polymers can retain a good balance of properties at a higher PVCH level.   It was hoped that making the butadiene block of high 1,2 microstructure in a 70SBS would give higher impact strength to the hydrogenated product, analogous to the rubberizing effect noted in the 25SB(1,2)S-R (Table IV, No. 7).   The impact was slightly higher (*cf.* Nos. 7 and 8), but stiffness was decreased.   The optimum balance of properties for an engineering plastic (heat distortion temperature, stiffness, and impact strength) is found in the SBS-R and SB^M-R polymers containing 70-75% PVCH blocks.   The heat resistance is not as sensitive to loading as in high melting crystalline hydrocarbon polymers like polypropylene and poly-4-methylpentene.   A comparison is made in Table VI.   The useful temperature at full loadings for the PVCH polymers is significantly higher than these crystalline hydrocarbon polymers.

Table VII.   Light Transmission through Transparent Plastics

Polymer	450 mμ, %T	550 mμ, %T	600 mμ, %T
PMMA	93	93	93
PS	88	89	90
PVCH	88	90	91
70SBS-R	75	79	82
25SB^M-R	72	77	79

The fact that the entire family of PVCH plastics is clear adds to the value of these polymers.   Data reflecting the level of light transmission are presented in Table VII.

PVCH is comparable with polystyrene in light transmission although both have transmission less than PMMA.   It should be possible to im-

Table VIII.   PVCH Blends with Hydrogenated SB^M Polymers

No.	Wt. % PVCH	Wt. % SB^M-R	Wt. % PVCH in SB^M-R	Flex. Mod., psi × 10³	Micro HDT, °C	Izod Impact, ft lbs/in	Appearance
1	90	10	25	—	—	—	opaque
2	85	15	25	—	—	—	transparent
3	90	10	70	—	—	—	transparent
4	75	25	25	210	126	0.3	transparent
5	70	30	25	120	127	0.9	transparent
6	65	35	25	90	120	1.6	transparent
7	60	40	25	71	108	3.7	transparent
8	10	90	56	90	125	no break	transparent

prove transparency in the hydrogenated block polymer by increasing the
1,2 content in the diene block. In their present form they both give
x-ray patterns characteristic of low density polyethylene. A higher 1,2
content in the starting diene block could eliminate the crystalline polyethylene contribution.

PVCH Blends With Hydrogenated SB Polymers. One early goal
of this research was to find a polymer that could be blended with PVCH
to give an impact resistant product. Conventional unsaturated rubbers
were incompatible. We did find, however, that PVCH-containing copolymers were compatible with PVCH and could yield clear, tough
plastic compositions as shown in Table VIII. To obtain transparency,
a minimum of 15% of a 25% PVCH compolymer is required. Depending
on the ratios of PVCH to the copolymer, tough heat-resistant products
can be made.

In summary the hydrogenation of styrene–diene block polymers is
a practical route to tough, clear heat-resistant plastics. Variations in
overall physical properties can be obtained by controlling composition,
microstructure, and the nature of the block sequence using conventional
anionic polymerization techniques.

Literature Cited

1. Fettes, E. M., "Chemical Reactions of Polymers," Interscience, New York,
    1964.
2. Staudinger, H., Fritschi, J., *Helv. Chim. Acta* (1922) **5**, 785.
3. Staudinger, H., *et. al.*, *Ber.* (1929) **62**, 263-267.
4. Fishcher, E., *Ber.* (1913) **46**, 3288.
5. Jones, R. V., *et al.*, *Ind. Eng. Chem.* (1953) **45**, 1117.
6. Moberly, C., *et. al.*, U.S. Patent **3,023,201** (1962).
7. Steinhofer, A., *et. al.*, German Patent **1,131,885** (1962).
8. Haefele, W., *et. al.*, U.S. Patent **3,333,024** (1967).
9. Hoeg, D. F., *et. al.*, U.S. Patent **3,598,886** (1971).
10. Moberly, C. W., "Encyclopedia of Polymer Science and Technology,"
    Vol. 7, p. 557 Interscience, New York, 1968.
11. Wicklatz, J., "Chemical Reactions of Polymers," E. M. Fettes, Ed., p. 173,
    Interscience, New York, 1964.
12. Parker, J., U.S. Patent **2,456,428** (1948).
13. Gluseshkamp, E., *et al.*, U.S. Patent **2,844,573** (1958).
14. Steinhofer, A., *et. al.*, Canadian Patent **718,089** (1965).
15. Warner, A., *et. al.*, U.S. Patent **2,726,233** (1955).
16. Hsieh, H. L., *Rubber Plastics Age* (1965) **46**, 394.
17. Kuntz, I., *J. Polymer Sci.* (1961) **54**, 569.
18. Gilman, H., *J. Amer. Chem. Soc.* (1944) **66**, 1515.
19. Holden, G., *et. al.*, U.S. Patent **3,265,765** (1962).
20. Chang, F., Goldberg, E., *Polymer Preprints* (1967) **8**, 39.

RECEIVED February 12, 1973.

$$4$$

Segmented Polyester Thermoplastic Elastomers

W. K. WITSIEPE [1]

E. I. du Pont de Nemours and Co., Inc., Elastomer Chemicals Department, Experimental Station, Wilmington, Del. 19898

The preparation and physical properties of a group of polyether-polyester copolymers are described. Where the polyester component has a high homopolymer melting point, the resulting copolyesters are tough, resilient, thermoprocessable elastomers. Polymer hardness and modulus characteristics may be varied from a fairly soft elastomer to an impact-resistant plastic by varying the relative amounts of polyether (soft segment) and polyester (hard segment). The general characteristics of these polymers suggest that these materials have a continuous amorphous polyether phase tied together with crystalline hard segment domains. Properties such as tear strength, flex resistance, and oil resistance may be modified by incorporating a second ester component; in these latter polymers, the crystalline phase contains one, but not both, of the hard segment ester components. Melting points of the copolymers depend upon the mole fraction of crystallizable hard segment.

This report deals with the synthesis and characterization of copolyesters derived from polyalkylene ether glycols, aromatic dicarboxylates, and short-chain aliphatic diols. Modification of polyethylene terephthalate by use of up to 20 wt % of high-molecular-weight polyethylene ether glycol was first reported independently by Coleman (1) and Snyder (2) to yield polymeric fibers having improved dye receptivity and moisture regain, as well as greater flexibility, compared with the unmodified polymer. Melting point and tenacity of the modified polymers were similar to those of unmodified polymers. Coleman (3) also reported that the polyether did not interfere with the crystallinity of polyethylene terephthalate, and that the polyether resided wholly in

[1] Present address: E. I. du Pont de Nemours and Co., Inc., Elastomer Chemicals Dept., P.O. Box 1378, Louisville, Ky. 40201

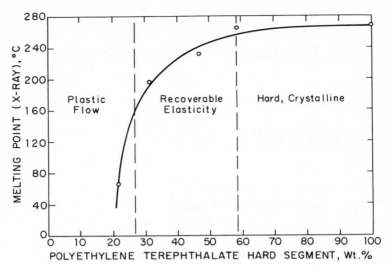

Figure 1. Melting point (x-ray) of segmented polyesters having a 4000 $\overline{M}_n$ polyether glycol soft segment and a polyethylene terephthalate hard segment (5)

amorphous regions of the semicrystalline fibers. Copolymers containing 30% polyether had limited elasticity.

J. C. Shivers (4, 5) broadened this approach to include other poly-alkylene ether glycols and polyester hard segments. Characterization of polymers having a much broader range of ether content showed that a wide variety of products—ranging from hard plastics to semicrystalline elastomers and, at high ether content, to soft elastic gums—could be obtained (Figure 1). The elastomers described by Shivers were reported to have excellent stress decay and tensile recovery when tested as drawn fibers. Subsequently, other workers investigated the use of polyether esters in various fiber, film, and adhesive applications (6, 7, 8, 9, 10). Certain characteristic properties of aromatic polyesters such as high melting point, insolubility in most solvents, excellent melt stability, and high strength make them of interest as hard segments in thermoprocessable elastomers.

Experimental

Catalyst. Magnesium metal (1.41 grams, 0.058 gram-atom) is added to 300 ml of dry 1-butanol and the butanol refluxed for about four hours in the absence of moisture. The magnesium reacts to form a gelatinous mass after; 36.0 grams (0.106 mole) of tetrabutyl titanate is then added and reflux continued for an additional hour. The resulting homogeneous solution is cooled and bottled until required.

Polymerization. These materials are placed in a 300-ml distillation flask fitted for distillation: polytetramethylene ether glycol (PTMEG), number-average molecular weight about 1000 (35.0 grams, 0.035 mole); 1,4-butanediol (4G) (25.0 grams, 0.28 mole); dimethyl terephthalate (DMT) (40.0 grams, 0.21 mole); *sym*-di-β-naphthyl-*p*-phenylenediamine (0.15 gram).

A stainless-steel stirrer with a paddle cut to conform with the internal radius of the flask is positioned about $\frac{1}{8}$ inch from the bottom of the flask and agitation is started. The flask is placed in an oil bath at 200°C, agitated for five minutes, and 0.3 ml of catalyst is added. Methanol distillation starts almost immediately, and distillation is practically complete in 20 minutes. The temperature of the oil bath is maintained for one hour after the addition of catalyst. The temperature of the bath is then increased to 260°C during about 30 minutes. The pressure on the system is then reduced to 0.5 mm Hg or less (about 0.1 mm Hg measured with a McLeod gauge at the pump) and distillation at reduced pressure is continued for about 90 minutes. The resulting viscous, molten product is scraped from the flask in a nitrogen (water- and oxygen-free) atmosphere and allowed to cool.

Materials. Except for the short-chain diols, all of the reagents were of commercial quality and used as received. The short-chain diols (Eastman) were distilled from a small quantity of sodium before use.

Test Methods. Inherent viscosities were determined at a concentration of 0.1 g/dl in *m*-cresol at 30°C and are reported in dl per gram.

Melting points and glass transition temperatures were determined in the usual way by use of a differential scanning calorimeter (Dupont Model 900). Heating rate was 11° C/min.

To determine mechanical properties, unless otherwise noted, all specimens were pressed at about 20°C above their melting point, held at this temperature, and then cooled under pressure over five to 10 minutes to room temperature. They were then conditioned at 24°C and 50% relative humidity for at least two days before testing.

Test methods used were:

Hardness, Shore	ASTM D2240
Tensile strength	ASTM D412
Elongation at break	ASTM D412, Die C
Tensile modulus	ASTM D412
Tear strength	ASTM D1938 at 50 in./min
Torsional modulus (Clash-Berg)	ASTM D1043
Compression set	ASTM D395–55, method B, 22 hr/70° C
Brittle point	ASTM D746

Discussion

The copolyesters may be considered as having been derived by randomly joining, head-to-tail, soft and hard segments. A generalized structure is:

$$\left[\ \left(\text{O--CH}_2\text{CH}_2\text{CH}_2\text{CH}_2\right)_x\text{OC--AR--C} \right]\ \left[\ \text{ODO--C--AR--C} \right]$$

Soft Segment   Hard Segment

AR = the aromatic moiety of the dicarboxylate
D = the alkylene portion of a short-chain diol
x = the number of tetramethylene ether units
in the polytetramethylene ether glycol

Copolyesters have been prepared from dimethyl terephthalate, poly-tetramethylene ether glycol (molecular weight, 1000) and various short-chain diols, especially 1,4-butanediol and ethylene glycol. With 1,4-butanediol, the crystalline hard segments (which produce the high-modulus tie-point domains) consist of consecutive units of tetrameth-ylene terephthalate (4GT); with ethylene glycol, these segments consist of ethylene terephthalate (2GT). The amorphous phase contains units of polytetramethylene ether glycol terephthalate (PTMEG-T).

Copolyesters containing two aromatic carboxylate residues have been synthesized, too. For example, 1,4-butanediol has been used with dimethyl terephthalate in combination with a number of dimethyl esters, including dimethyl phthalate (4GP), dimethyl isophthalate (4GI), di-methyl sebacate (4G10), and dimethyl m-terphenyl-4,4″-dicarboxylate (4GTP).

For convenience, polymer compositions will be specified according to their hard segment weight percentage and soft segment composition. For example, 44% 4GT/PTMEG-T is a random segmented copolymer containing 44% by weight of tetramethylene terephthalate segments and 56% by weight of polytetramethylene ether terephthalate. Unless other-wise specified, all data refer to polytetramethylene ether glycol having a number-average molecular weight of 980.

Synthesis

The polyether esters are made by typical melt polymerization pro-cedures. A prepolymer is first prepared by interchange of the methyl ester of one or more aromatic dicarboxylic acids with a mixture of a polymeric diol and enough short-chain diol for an overall 50% excess of hydroxyl functionality (Figure 2). A titanate catalyst is generally used. Methanol is fractionated from the reaction mixture to avoid loss of

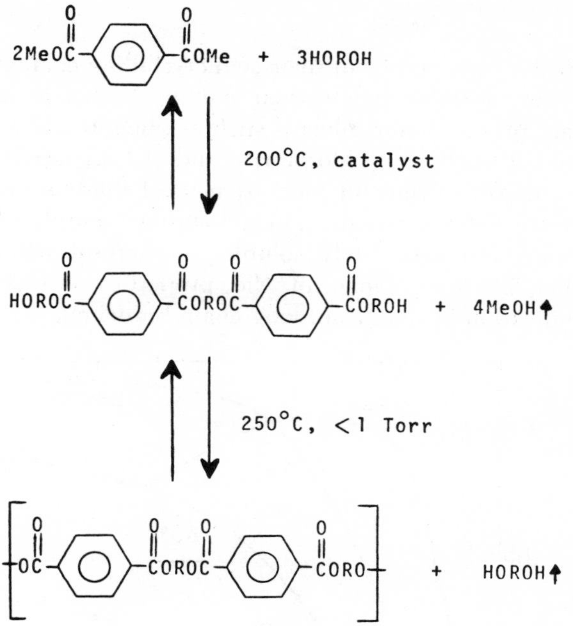

*Figure 2.   General method for synthesizing seg-
mented polyester elastomers.   HOROH is a mixture
of a long-chain diol and a short-chain diol.*

either methyl esters or short-chain diol.   The ester prepolymer is driven
to high molecular weight by distillation of the original short-chain diol
excess at 240°-250°C at less than 1 mm.   The course of the reaction may
be followed by measuring the power required to stir the reaction mass
at some slow, constant rate.   This rate will plateau after about 90
minutes, depending upon precise conditions of agitation, temperature,
and pressure.   Final polymer molecular weight is determined by the
rate of competing degradation and coupling reactions.   As Figure 3
shows, the polymerization rate increases as reaction temperature increases.   However, at a high reaction temperature of 265°C, final molecular weight is limited by the onset of degradation.   Optimum reaction
temperature is about 250°C.   At this temperature, degradation is slow,
and final molecular weight once achieved is fairly insensitive to continued reaction time.

Although this paper is concerned primarily with polytetramethylene
ether glycol as the polymeric component of the copolyester elastomer,
other polyglycols can be used; for example, polyethylene ether glycol
and poly-1,2-propylene ether glycol.   Various combinations of short-chain
diols and dicarboxylates may be used also, as mentioned (4).

Polymer Characterization

As might be expected from their semicrystalline character, the poly-ether esters are insoluble below their melting points in most solvents. Exceptions are proton donor solvents such as phenols and a few partially chlorinated hydrocarbons (chloroform and 1,1,2,2-tetrachloroethane). m-Cresol is a useful solvent for measurement of dilute-solution viscosity, osmotic-pressure determinations, and gel-chromatographic fractionation. The copolymers are completely soluble in m-cresol without giving a significant gel fraction. None of the properties evaluated thus far shows any indication of significant long-chain branching.

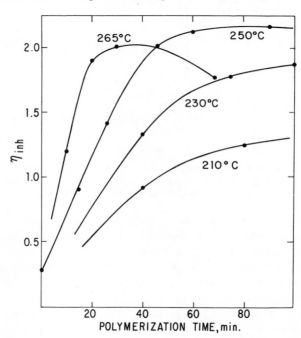

Figure 3.   Polymerization of a 58% 4GT/PTMEG-T copolymer   at   0.03   torr.   Catalyst   is   0.22mM Ti(OC₄H₉)₄ and 0.13mM Mg(CH₃COO)₂ per 100 g of polymer.

Ultimate properties such as tensile and tear strength depend upon polymer molecular weight. Stress-strain curves for three polymers hav-ing the same composition but different inherent viscosities are compared in Figure 4. All three polymers have similar moduli at low elongations, but they differ in modulus at high extension and elongation at break. Mechanical properties are relatively unaffected by molecular weight after a certain minimum molecular-weight level is reached. Although they vary with composition, optimum properties are usually obtained at molecular weights of about 25,000, corresponding to an inherent viscosity of about 1.5.   The polymer has the expected geometric molecular-weight distribution (Figure 5).

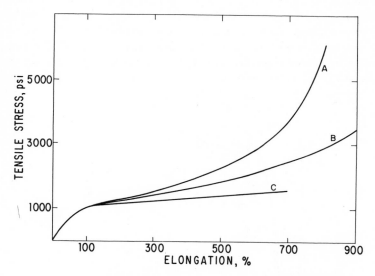

Figure 4.   Stress-strain diagrams of a 33% 4GT/PTMEG-T copolymer; A, η_{inh} = 1.97; B, η_{inh} = 1.40; C, η_{inh} = 0.99

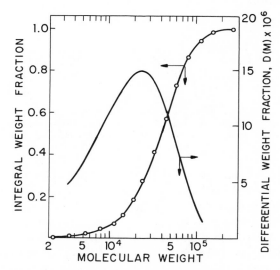

Figure 5.   Molecular-weight distribution of a 58% 4GT/PTMEG-T copolyester by gel permeation chromatography; polymer η_{inh} is 1.49

Calorimetric data provide evidence that the terpolymers contain a sizable crystalline fraction. Figure 6 compares the DSC trace of a physical mixture of high-molecular weight 4GT and PTMEG-T (ether linkage molecular weight is 2000) with that of a copolymer having the same overall composition. The position of the 4GT melting endotherm in the mixture is about 2°C lower than that of pure 4GT. However, the corresponding melting point in the ether-ester copolymer of the same overall composition is shifted downward about 47°C. Similarly, the endotherm corresponding to that of the soft segment homopolymer is about 12°C lower in the copolymer.

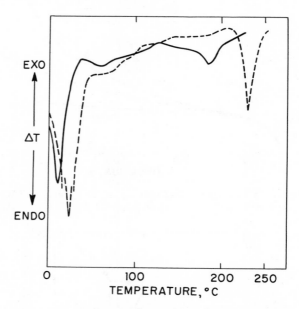

Figure 6. Differential scanning calorimeter traces of a 33% 4GT/PTMEG(2000)-T copolymer and of a polymer blend having the same composition; solid line represents copolymer; dashed line represents blend.

Copolymer melting points depend generally upon the mole fraction of 4GT, as suggested by Flory (11). Figure 7 shows the melting points of various polymers (as their reciprocals in °K) plotted against the mole fraction of 4GT present. On a molar basis, the polyether and various short-chain ester segments are about equally effective in reducing copolymer melting point despite the wide differences in their respective molecular weights. It is apparent that a given weight fraction of a high-molecular-weight soft-segment comonomer will result in less melting point depression than will a lower-molecular-weight comonomer

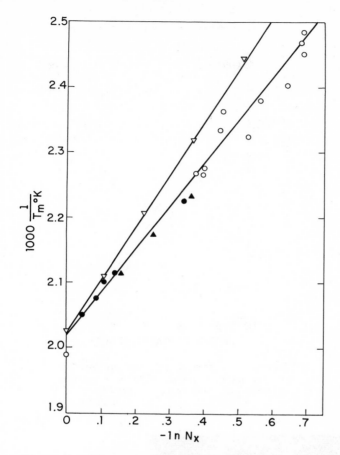

Figure 7.   Effect of 4GT mole fraction (N$_x$) on the melting point (T$_m$ by DSC) of various copolyesters; ● = *4GT/PTMEG-T;* ▲ = *4GT/4GI;* ○ = *4GT/ 4GI/PTMEG-T;* △ = *4GT/4G-10; data of Ueberreiter and Steiner (12)*

(Figure 8).   Since a relatively large weight fraction of soft segment must be incorporated into 4GT to obtain products having a low enough modulus for many elastomeric applications, the use of a high-molecular-weight soft segment is mandatory if adequate heat resistance is to be obtained in these random copolymers.

X-ray measurements confirm and extend the calorimetric results. A diffraction pattern of a drawn fiber of a 4GT/PTMEG-T copolymer having 57.5% by weight of 4GT is shown in Figure 9.   The patterns are

crystallites are the same in both the copolymer and homopolymer.   This identical with that of 4GT homopolymer, except for the presence of an amorphous halo in the copolymer.   Crystal spacings of hard-segment

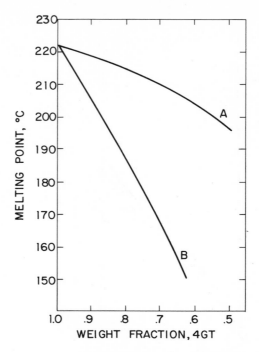

Figure 8.   Effect of 4GT weight fraction on the melting points (DSC) of various copolyesters; calculated from Figure 7; A = 4GT/PTMEG-T; B = 4GT/4G10

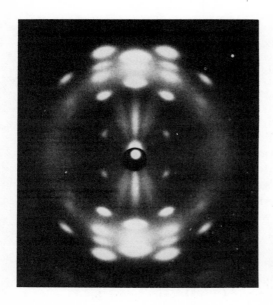

Figure 9.   Wide-angle x-ray diffraction pattern of a stretched fiber of a 57% 4GT/PTMEG-T copolyester

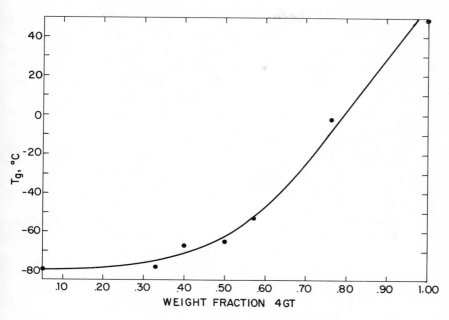

Figure 10.   Glass transition temperature of 4GT/PTMEG-T copolymers

is interpreted to mean that the crystallized phase is essentially free of soft-segment material.

Glass-transition temperatures of 4GT/PTMEG-T copolymers, as measured by DSC, are in accord with their structures as suggested by the foregoing calorimetric and x-ray studies. The soft segment T_g is close to that reported (*13*) for polytetramethylene ether glycol ($-84°C$) when the hard segment content is below about 40% (Figure 10). The DSC does not show a clearly defined hard segment T_g, but curvature of the scan at about 50°C could be interpreted either as a T_g or as a minor endotherm. At very high hard-segment content—for example, 76% 4GT—an apparent soft-segment T_g occurs at $-2°C$. These changes can be rationalized on the basis of more 4GT segments being forced into the amorphous region at the higher hard-segment level. In addition, the degree of pseudocross-linking has been increased. This effectively decreases the length of unrestricted soft-segment polytetramethylene ether chains. Copolymer T_g is not reflected in low-temperature properties such as brittle point. For example, the 76% 4GT/PTMEG-T copolymer referred to has a brittle point of $-60°C$.

Based on these data and on electron microscope examination, R. J. Cella (*14*) has concluded that phase separation occurs in polyether esters below their melting points, as shown in Figue 11. The continuous phase consists of essentially all of the soft-segment material, in addition to any polyester segments that are either too short to crystallize or cannot do so because of chain entanglement or high melt viscosity. The remaining polyester seems to be dispersed in the form of fibrillar crystalline lamellae closely connected at growth faces by short tie molecules. In addition, other ties to the crystalline domains serve to fix the amorphous phase material into an elastic network. In essence, then, a more-or-less continuous crystalline network is superimposed on a continuous amorphous network.

This morphology is partly inferred from the stress-strain curve of these materials (Figure 12). The curve is typically composed of three clearly defined regions. At low elongations, 5-10%, force increases almost linearly with elongation and deformation is reversible. Between about 50 and 200% elongation, force is relatively independent of further elongation and deformation is only partly reversible. At higher elonga-t·.ns, force increases with extension about as expected for a crystalliz-

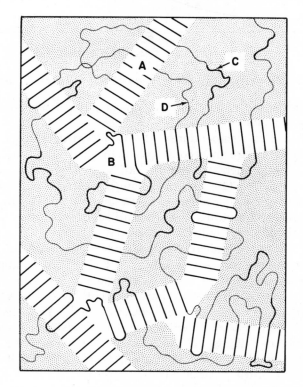

Figure 11. Schematic diagram of the proposed morphology of polyether ester copolymers; A = crystalline domain; B = junction area of crystalline lamella; C = polymer hard segment that has not crystallized; D = polymer soft segment

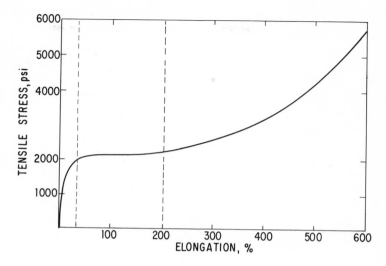

Figure 12. *Stress-strain curve of a 58% 4GT/PTMEG-T co-polyester. ASTM D-412 dumbells were strained at 2 in./min.*

able amorphous network. It appears that initial extension is due to reversible deformation of the crystalline domains and the short tie molecules that connect them. In the second area, reorientation of crystalline domains in the direction of stress allows plastic flow of both networks. At higher elongation, an increasing share of the force is borne by entropic extension of the original amorphous phase with relatively little further chain slippage.

Polymer Mechanical Properties

Mechanical properties depend upon the chemical structure of both the hard- and soft-segment constituents. Properties such as melting point, modulus, solubility, and resistance to creep and set are influenced primarily by the nature of the hard segment, its concentration, and its state of crystallization. Low-temperature flexibility as well as tensile and tear strength also vary with hard-segment composition and concentration but, in addition, strongly depend on the composition of the amorphous phase.

Copolymers having PTMEG-T soft segments and hard segments consisting of the terephthalates of a series of α,ω-diols exhibit a regular variation in modulus that correlates with hard-segment homopolymer melting point (Figure 13). At the levels of hard-segment content investigated, the lower moduli are associated with the lower melting points.

Modulus values vary regularly with hard-segment content (Figure 14). The soft-segment homopolymer itself, PTMEG-T, is a soft gum that flows readily when stressed. Depending upon the molecular weight of the polyether moiety, copolymers having as little as 10-20% 4GT hard segment resist permanent deformation, but have relatively low strength at 25°C, with marked loss of strength at 70°C. Polymers having more than about 30% 4GT hard segment are tough and retain their strength at elevated temperature.

Unlike modulus, tear resistance is not clearly related to hard-segment melting point. For example, polymers having a trimethylene terephthalate hard segment have considerably lower tear resistance than those having a 4GT hard segment, although the homopolymer melting points are comparable. With a given hard segment—4GT, for example—

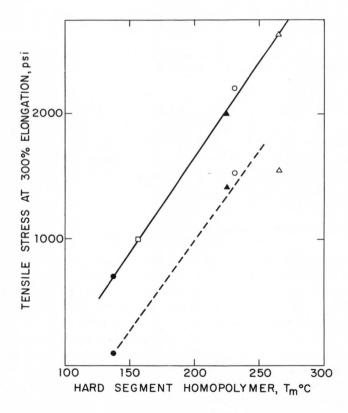

Figure 13. Variation of tensile stress (300%) with hard-segment melting point in PTMEG-T copoly-esters; △ = 2GT; ▲ = 3GT; ○ = 4GT; ● = 5GT; □ = 6GT; _ _ _ _ _ 32% hard segment; _____ 50% hard segment

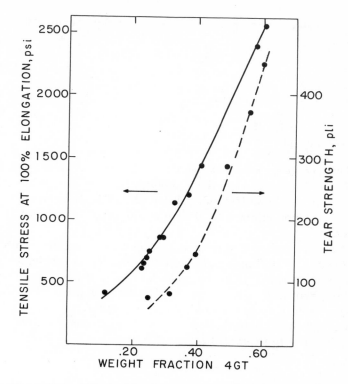

Figure 14. Effect of hard-segment weight fraction on the tensile stress at 100% elongation and tear strength of 4GT/PTMEG-T copolymers; tensile stress is represented by solid line, tear strength by dashed line

trouser tear resistance at 50 inch per minute depends upon hard segment content in much the same way as does modulus (Figure 14).

At a given hard-segment content, tear resistance is more readily correlated by amorphous phase composition. For example, the tear resistance and tensile strength of polymers made using a polypropylene ether-terephthalate soft-segment glycol in place of PTMEG-T are relatively low (Table I). The changes are in the direction expected from results reported using the two polyols in various polyurethanes (*15*). These effects are believed to result from the tendency of polytetramethylene ether sequences to stress-crystallize at high elongation.

The amorphous phase composition may also be modified by use of mixed short-chain ester segments. Partial replacement of the 4GT hard segment with another results both in decreasing 4GT concentration and in reducing its overall mole fraction. The effects on modulus are shown in Figure 15 for the PTMEG-T/4GT/4GI system. As the terephthalate mole fraction is decreased in a polymer having constant total

Table I. Comparison of Polyether Soft Segments

Polyol composition	*PTMEG*	*PPG*
Polyol $\overline{M}_n$	1990	1940
η_{inh}, dl cresol/g	1.56	1.28
Stress at 100% elongation, psi	2020	2010
Tensile strength, psi	7700	3400
Elongation at break, %	640	470
Tear resistance, pli	300	130

"phthalate" content, modulus decreases until, at a 4GT mole fraction of 0.25, the system is essentially amorphous and the polymer has no strength. As the mole fraction of terephthalate is decreased further, 4GI crystallization apparently becomes important and modulus again increases.

These effects are not unique to copolymers in the isophthalate/terephthalate system but are the expected results of copolymerizing the crystallizable tetramethylene terephthalate hard segment with any relatively lower-melting short-chain diol ester segment. The apparent eutectic at a 4GT mole fraction of 0.25 represents the condition where there are few short-chain diol ester segments present of either type having sufficient length to crystallize.

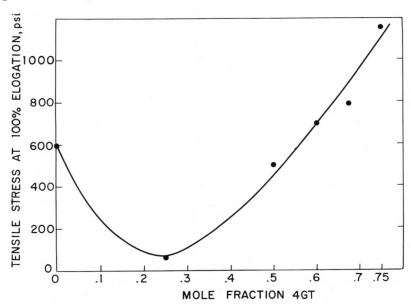

Figure 15. Effect of 4GI mole fraction on the tensile stress at 100% elongation of 4GT/4GI/PTMEG-T/I copolyesters; 4GI + 4GT content is constant at 33 wt %

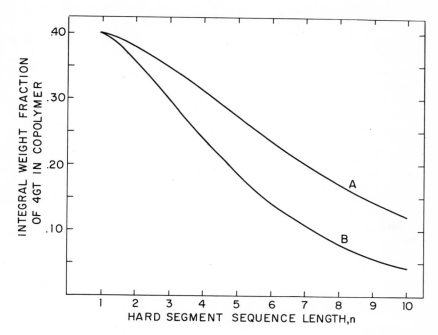

Figure 16. Weight distribution of hard-segment sequences in polyester copolymers; 40% 4GT/PTMEG-T (A), 40% 4GT/10% 4G-P/PTMEG-T/P (B)

If we assume a random distribution of hard segment units along the polymer chain, then the weight fraction of hard segment units having a given length in the copolymer may be calculated from the overall composition. Two such calculations are shown in Figure 16 for a 40% 4GT/PTMEG-T copolymer and for a similar copolymer in which 10% of the PTMEG-T is replaced with an equal amount of tetramethylene phthalate. In the case of the polymer that contains only terephthalate, the distribution of 4GT in sequences of varying length is relatively broad, and 31% of the copolymer consists of 4GT in sequences at least four units in length. However, the phthalate copolymer contains only 24% 4GT in sequences of the same length, although the overall terephthalate contents of the two polymers are the same. The mole fraction of phthalate in the mixed-acid copolymer is so low that only very short noncrystallizable sequences could be expected. This difference in polymer structure results in a decrease in the content of high-melting crystalline domains in the copolymer, and introduces noncrystalline esters into the amorphous phase.

Some of the effects of using a combination of short chain ester units are shown in Table II (*16*). Modulus at 100% elongation appears to be

Table II. Physical Properties of PTMEG-T/4GT Polymers Containing Second Short-Chain Ester Group

4GT Weight fraction	0.40	0.40	0.40	0.40	0.50	0.40
Other short-chain ester, weight fraction	0.1	0.10	0.10	0.10	—	—
Other short-chain ester, structure	4G10	4GP	4GI	4GTP	—	—
η_{inh}, dl cresol/g	1.6	1.7	1.8	1.9	1.6	1.6
Stress at 100% elongation, psi	1040	1040	1100	1200	1800	1210
Stress at 300% elongation, psi	1250	1320	1420	1720	2200	1600
Tensile strength, psi	4480	6670	7440	7350	7400	2500
Elongation, %	860	760	780	520	780	730
Tear resistance, pli	150	190	220	650	290	150
Clash Berg, $T_{10,000}$	−53	−46	−44	−10	−26	−52
Shore D hardness	41	43	40	49	49	44
Compression set, 22 hr./70° C, %	54	45	70	80	54	53

more nearly a function of the 4GT content and is relatively independent of the other added short-chain ester. Even when the added short-chain ester is very high-melting—for example, tetramethylene 4,4″-terphenyl dicarboxylate—the modulus is not increased as it is in a copolymer having only 4GT hard segments with the same total content of short-chain ester. However, other properties (such as tear resistance and low-temperature flexibility) tend to depend more on amorphous phase composition. The addition of small amounts of phthalate, isophthalate, and m-terphenyl-4,4′-dicarboxylate short-chain ester units thus generally increases tear resistance without a commensurate increase in either polymer hardness or 100% elongation modulus. In addition, low-temperature flexibility is adversely affected when the added ester is one that has a relatively high glass-transition temperature. These changes are reflected in other types of rupture properties, such as abrasion resistance and flex resistance.

The tensile strength of a polymer containing an additional noncrystalline short-chain ester segment tends to reach a high value at relatively low inherent viscosity than one having only 4GT segments. Reduced volume swell in contact with nonpolar oils and fuels may also be attributed to the greater ester content of the amorphous phase.

To a degree, the mechanical properties of copolyether esters depend upon the conditions under which the sample is allowed to crystallize. Polymers having 4GT hard segments crystallize very rapidly compared with some other terephthalate copolymers. This is apparent from the rate of hardness development of injection-molded samples. Figure 17 compares the hardness of injection moldings at a 57% 4GT/PTMEG-T copolymer with that of a 2GT copolymer at various times after ejection of the molding.   The hardness of the 4GT copolymer reaches its final value almost immediately, while the hardness of the 2GT polymer is still increasing after 36 hours.

The mechanical properties of polymer samples that crystallize slowly depend markedly on precise molding conditions as well as upon heat history after the molding operation.   In general, annealing for several hours at 60°-90°C below the polymer melting point will lead to improved compression set and resistance to creep, and to decreased hardness and tangent modulus.   Tensile stress at elongations greater than 100% increases.   The annealing operation does not affect polymer inherent viscosity.

Similar though less marked changes occur in the mechanical properties of fast-crystallizing polymers, such as those having 4GT hard segments.   The most notable change is in compression set.   The optimum temperature for improvement of compression set is about 70°-80°C below the polymer's melting point (Figure 18).

Lack of recovery after compression loading at elevated temperature probably results from a combination of creep and recrystallization of the sample while in the stressed condition.   That recrystallization plays a

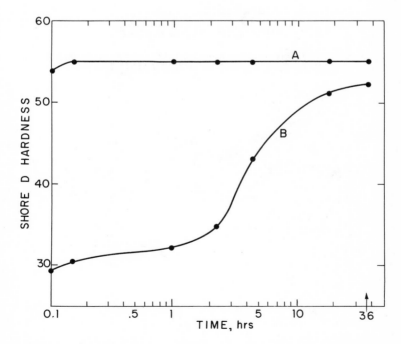

Figure 17.   Hardening rate of injection-molded samples of co-polyether esters.   Melt temperature is 240°C; mold temperature is 30°C; 58% 4GT/PTMEG-T (A), 56% 2GT/PTMEG-T (B)

major role is apparent from changes in DSC traces after the annealing operation. An unannealed 28% 4GT/PTMEG-T sample has broad melting endotherms centered around 63° and 165°C (Figure 19a). After

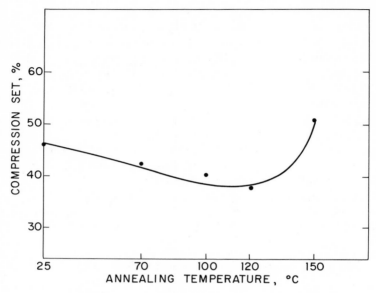

Figure 18. *Effect of annealing (24 hours) on the compression set (22 hr/70°C) of a 57% 4GT/PTMEG-T copolymer*

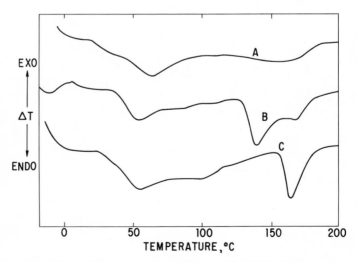

Figure 19. *DSC traces of a 28% 4GT/PTMEG-T copolymer as molded (A), after annealing 4 hours at 100°C (B), and after annealing 4 hours at 150°C (C)*

annealing for four hours at 100°C (Figure 19b), both endotherms are sharpened and enlarged. This recrystallization presumably will occur in a direction that tends to relieve stress in the polymer sample. When the same polymer was annealed at 150°C (Figure 19c), the endotherm at 165°C was quite sharp but relatively small. The low-melting endotherm is broad and, presumably, the polymer would recrystallize during compression-set testing at 70°C. The new network formed by the recrystallization process would tend to prevent sample recovery after release from the stressed condition.

All of these changes after annealing are consistent with the assumption that the amorphous region of the unannealed sample contains a considerable content of uncrystallized hard segment. The annealing process redistributes this into the hard phase and thus forms a more stable network.

Conclusions

The copolyester elastomers exhibit the properties expected for polymer systems in which a continuous amorphous phase is crosslinked by the formation of high-modulus domains. The general properties of the materials, the response of the system to copolymerization, and x-ray diffraction and thermal data show that network formation is basically a result of partial crystallization of the copolymer. This contrasts with the styrene/butadiene triblock polymers, where phase separation results from insolubility of a glass-forming polymer block in the continuous phase (17), and with polyurethanes, where domain formation is thought to result from the presence of either strong hydrogen bonding or, in some instances, microcrystallinity (18, 19).

Acknowledgments

The collaboration of L. Plummer and J. R. Wolfe in polymer structure property relationships, of G. K. Hoeschele (20) in polymerization studies, and of R. E. Fuller in polymer fractionation is gratefully acknowledged. Polymer morphology investigations were done by R. J. Cella. End-use applications have been investigated by M. Brown (21) and D. Bianca.

Literature Cited

1. Coleman, D., British Patent **682,866** (1952).
2. Snyder, M. D., U.S. Patent **2,744,087** (1953).
3. Coleman, D., *J. Polym. Sci.* (1960), **14**, 15.

4. Shivers, J. C., U.S. Patent **3,023,192** (1962).
5. Charch, W. H., Shivers, J. C., *Text. Res. J.* (1959), **29**, 536.
6. Bell, A., Kibler, C. J., Smith, J. G., U.S. Patent **3,277,060** (1966).
7. Kobayshi, H., Japanese Patent application **44–20469/1969**.
8. Michel, R. H., U.S. Patent **2,865,891** (1958).
9. Nishimura, A. A., Komagata, H., *J. Macromol. Sci.* (1967), **A-1**, 619.
10. Willard, A., U.S. Patent **3,013,914** (1961).
11. Flory, P. J., *J. Chem. Phys.* (1949), **17**, 223.
12. Ueberreiter, K., Steiner, N., *Makromol. Chem.* (1964), **74**, 158.
13. Wetton, R. E., Allen, G., *Polymer* (1966), **7**, 356.
14. Cella, R. J., IUPAC meeting, Helsinki, 1972.
15. Berger, S. E., Szukiewicz, W., *Rubber Chem. and Tech.* (1965), **38**, 150.
16. Wolfe, J. R., private communication.
17. Holden, G., Bishop, E. T., Legge, N. R., *J. Polym. Sci, Part C* (1969), **26**, 37.
18. Bonart, R., *J. Macromol. Sci. B 2* (1968), 115.
19. Seymour, R. W., Estes, G. M., Cooper, S. L., *Macromolecules* (1970), **3**, 579.
20. Hoeschele, G. K., Witsiepe, W. K., *Angew, Makromol. Chem.*, in press.
21. Brown, M., Witsiepe, W. K., *Rubber Age* (N.Y.) (1972), **104**(3), 35.

RECEIVED April 13, 1972.

Synthesis and Properties of Imide and Isocyanurate-Linked Fluorocarbon Polymers

J. A. WEBSTER, J. M. BUTLER, and T. J. MORROW

Monsanto Research Corp., 1515 Nicholas Rd., Dayton, Ohio 45407

Sulfur tetrafluoride fluorination of aryl perfluoroalkyl esters provides a route to stable aryl perfluoroalkyl ether compounds (1). By using nitrophenyl esters of perfluoroalkylene and perfluoroalkylene ether dicarboxylic acids, α, ω-bis (m-nitrophenoxy) perfluoroalkylene ether intermediates were prepared. The conversion of these products to the corresponding amino and isocyanatophenoxy derivatives is described. Reaction of the diamines with benzophenonetetracarboxylic dianhydride resulted in formation of polyimides. Cyclotrimerization of the diisocyanate intermediates formed polyisocyanurates. Both the imide and isocyanurate polymers have high thermal, oxidative, and hydrolytic stability.

The high chemical and thermal stability of structures containing perfluoroalkylene, $(CF_2)_n$, and perfluoroalkylene ether, $[R_fO]_n$ make chain segments of this type desirable building blocks for high stability polymers. Various addition and condensation polymers containing these segments have been reported. Amidine- and nitrile-terminated perfluoroalkylene ether derivatives, for example, have been polymerized to form elastomeric triazine-linked polymers. These polymers have been studied extensively (2, 3). They show high thermal and oxidative stability although the fluorocarbon linked directly to the triazine ring has shown evidence of inducing hydrolytic instability (4). Chain segments of perfluoroalkylene and perfluoroalkylene ether have also been linked through siloxane and urethane moieties (5, 6). In these polymers, separation of the fluorocarbon segment from the linking moiety by one or more methylene units has been used to gain stability. The presence of aliphatic hydrocarbon moieties introduces potential sites of oxidative instability, however.

Critchley *et al.* (7) prepared high stability amide- and imide-linked fluorocarbon polymers from perfluoroalkylene diamine intermediates of the type:

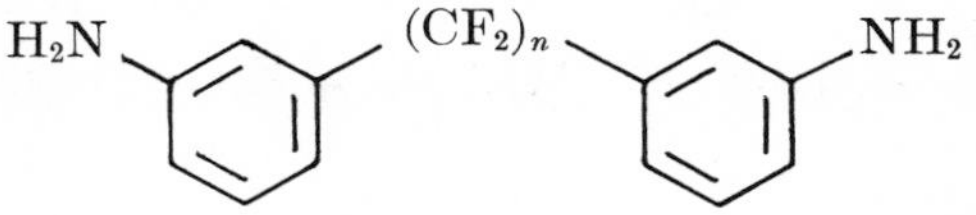

where $n=3$.

In these polymers the reactive functional group and the fluorocarbon segments are linked by thermally and oxidatively stable phenylene groups. More recently, Yakubovich *et al.* (8) and Feast and co-workers (9) reported similar fluorocarbon polymers, in which the phenylene moiety separates the fluorocarbon segment from the functional group involved in the polymer-linking reaction.

For the full, high-stability potential of the fluorocarbon chain segments to be realized in the development of new polymers, the chemical bonds linking the fluorocarbon segments must provide an optimum combination of thermal, oxidative, and hydrolytic stability. It is toward this end that the development of appropriate aryl-terminated fluorocarbon intermediates has been sought. The reactive substituents on the aryl rings have included amine and isocyanate groups capable of forming stable imide and isocyanurate polymers. The synthesis of intermediates and the polymers, together with the properties of representative polymers, are described here.

Perfluorinated Intermediates

The available sources of perfluoroalkylene segments are limited to perfluoroalkylene diiodides and dicarboxylic acids of the type $I(CF_2CF_2)_nI$, where $n=1$ to 5, and $HOOC(CF_2)_nCOOH$, where $n=$

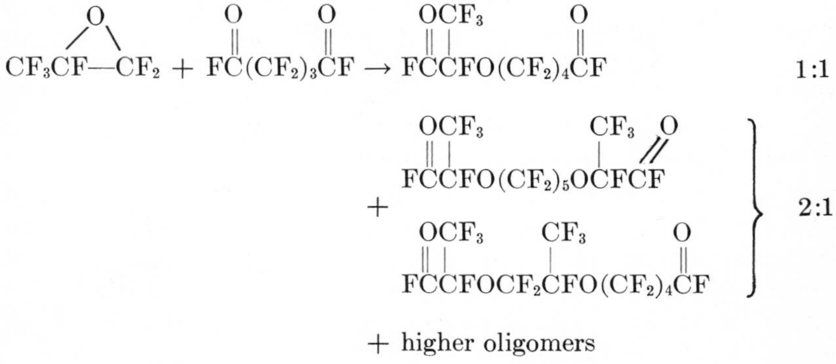

+ higher oligomers

3 to 8.   Perfluoroalkylene ether chain segments are also available as the oligomer dicarboxylic acids formed by the addition of hexafluoro-propylene epoxide to perfluoroglutaryl fluoride, as described by Fritz and Warnell (*10*).

Linear perfluoroalkylene ether dicarboxylic acids, HOOC-$(CF_2CF_2OCF_2CF_2)_nCOOH$, are reported to have been prepared by UV-catalyzed coupling reactions (*6*) and oxidative polymerization reactions (*11*).   Although these structures are considered to represent the pre-ferred chain segments, the synthesis methods apparently have not been developed enough for commercial availability.

Synthesis of Intermediates

Two known synthesis routes were used to convert short-chained perfluoroalkylene dicarboxylic acid and diiodide intermediates to diaryl intermediates.   The copper-catalyzed coupling of aryl and fluorocarbon iodides described by McLoughlin and Thrower (*12*) provided con-venient synthesis of short-chained diphenylperfluoroalkylene compounds when appropriate fluorocarbon diiodides were available.   The prepara-tion of diaryl intermediates *via* aryl ketone synthesis and subsequent SF_4 fluorination reaction provided a satisfactory alternative synthesis of lower-molecular-weight intermediates (*13*).

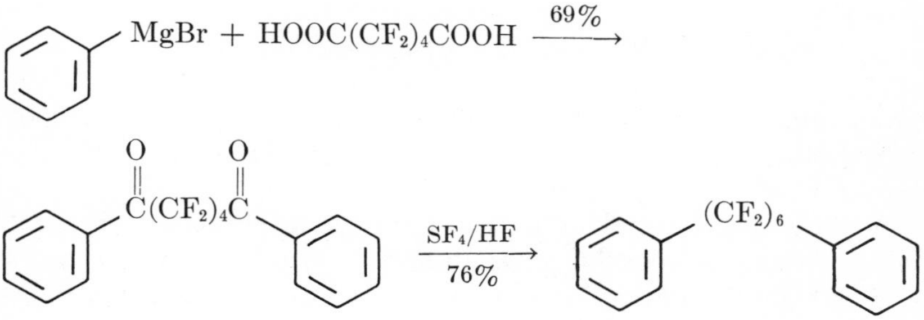

SF_4 fluorination of the diketone from perfluoroglutaric acid, however, formed a high yield of the cyclic ether.

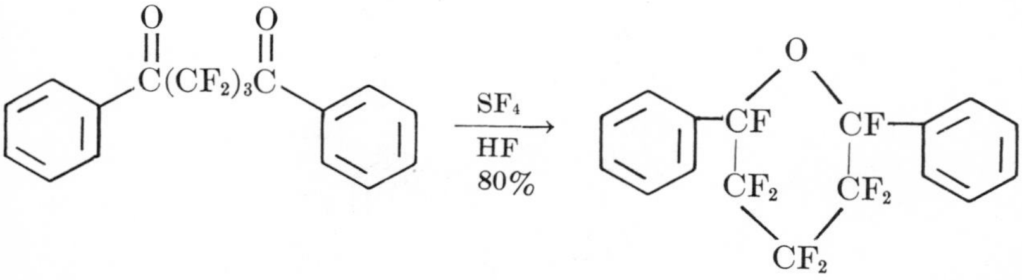

An interesting observation was that this diketone is a relatively strong acid and, when titrated with 0.1N NaOH, required only one equivalent/ mole, apparently forming the cyclic ether salt, I.

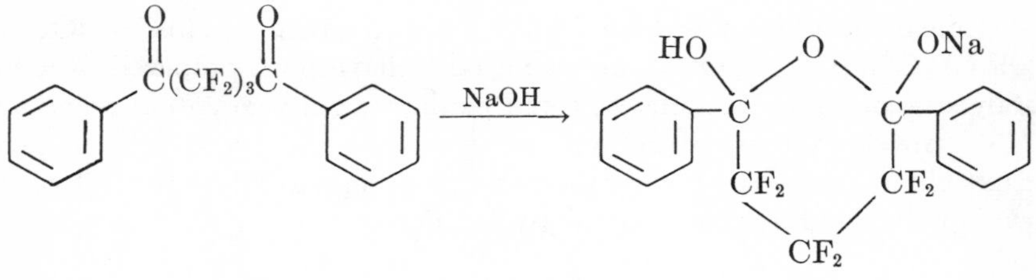

I

Acidification formed the diol II, which was isolated in 92% yield.

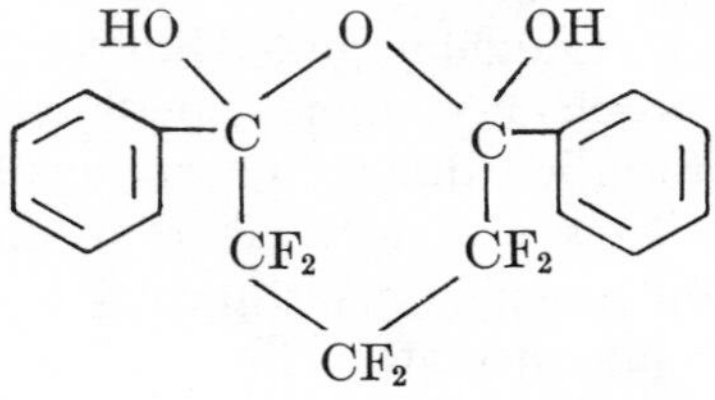

II

mp, 89°–91°C

Attempts to form the aryl diketone products from higher perfluoro-alkylene ether dicarboxylic acids by the acylation or Grignard reactions, however, were relatively unsuccessful.

A third route, the fluorination of nitrophenyl esters of perfluoro-alkylene ether dicarboxylic acids, has now been developed for synthesis of highly stable aryl perfluoroalkylene ether intermediates of the type shown by III.

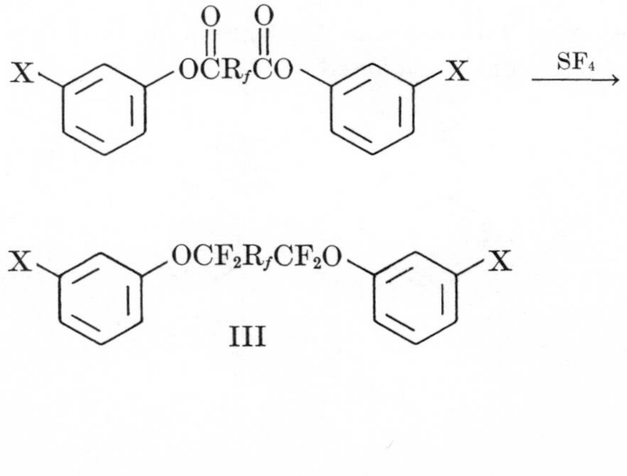

III

This method is an extension of the monoester fluorination reactions reported earlier by Sheppard (*1*). Initial effort was placed on developing conditions that would result in high yields in the fluorination of the nitrophenyl ester of perfluoroglutaric acid. This was followed by fluorination of the nitrophenyl esters of the higher 1:1, 2:1, and 3:1 fluorocarbon ether oligomer acids. The products were the corresponding nitrophenoxy derivatives represented by **IV** where $x + y = 1$, 2, and 3.

$$O_2N\text{—}\underset{}{\bigcirc}\text{—}O(CF_2\overset{\overset{CF_3}{|}}{C}FO)_x(CF_2)_5(O\overset{\overset{CF_3}{|}}{C}FCF_2)_y\,O\text{—}\underset{}{\bigcirc}\text{—}NO_2$$

IV

Catalytic reduction of these products formed the corresponding diamines. The simplest diamine—that is $x + y = 0$—was also converted to the corresponding diisocyanate in high yield by phosgenation.

Isocyanurate Polymers

Isocyanurate-linked polymers can be formed by cyclotrimerization of diisocyanate intermediates under moderate temperature conditions (*14*). Compounds V and VI cyclotrimerize readily at room temperature upon addition of catalytic quantities of *N,N,N',N'*-tetramethyl-1,3-butanediamine and allyl glycidyl ether, provided that atmospheric moisture is rigorously excluded.

$$OCN\text{—}\underset{}{\bigcirc}\text{—}(CF_2)_6\text{—}\underset{}{\bigcirc}\text{—}NCO$$

V

$$OCN\text{—}\underset{}{\bigcirc}\text{—}O(CF_2)_5O\text{—}\underset{}{\bigcirc}\text{—}NCO$$

VI

Elevated temperatures (about 100°C) are required to ensure complete utilization of the —NCO functional group (*see* reaction, p. 66).

Polymerization of V at its melting point (about 75°C) was too rapid to permit adding and mixing of the catalyst. Addition of a reactive monofunctional liquid diluent such as perfluorobutylphenyl isocyanate allowed the resulting isocyanate solution to be cooled to 40°C; the catalyst was then added and mixed before appreciable polymerization occurred. The monoisocyanate addition also served to decrease the degree of crosslinking.

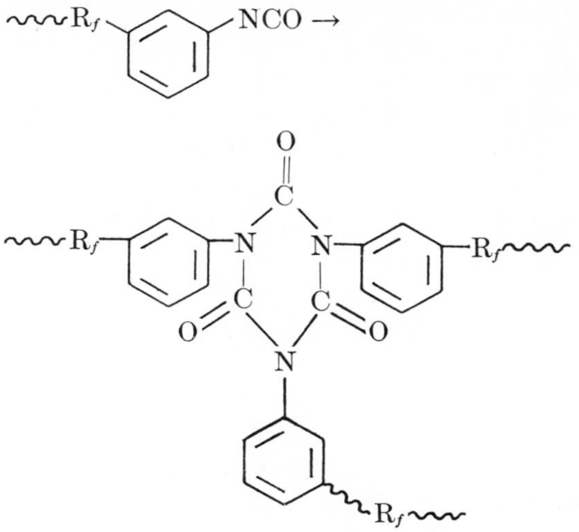

The polymers formed from these short chained intermediates were, as expected, hard, glassy solids. High stability of the polymers in air was shown by thermogravimetric analyses (Figure 1).

The high thermal stability of the isocyanurate structures was also shown by preparation of the cyclic trimer compounds prepared from m-perfluorobutyl and m-perfluorobutoxyphenyl isocyanates and determination of their thermal stability. The respective thermal stability values of the two purified isocyanurate compounds, as determined by the isoteniscope method of Blake et al. (15), were 373° and 370°C. Initial tests of hydrolytic stability of the polymers showed evidence of degradation after three days at 95°C and 95% relative humidity. Subsequent work showed susceptibility to hydrolysis was partially the result of incomplete polymerization. Properly cured polymer prepared from the phenyl–oxygen–fluorocarbon-containing diisocyanate showed no evidence of hydrolysis after a five-week exposure to conditions of 95% relative humidity at 95°C. A polymer specimen having the phenyl–CF_2 linkage showed evidence of hydrolysis after two weeks under the same conditions. These tests suggest greater stability of the phenoxy-linked structure.

Polymide Polymers

Polymide polymers were prepared by reaction of benzophenone-tetracarboxylic dianhydride with each of the four diamines having structure VII, where $(x+y) = 0, 1, 2,$ and 3.

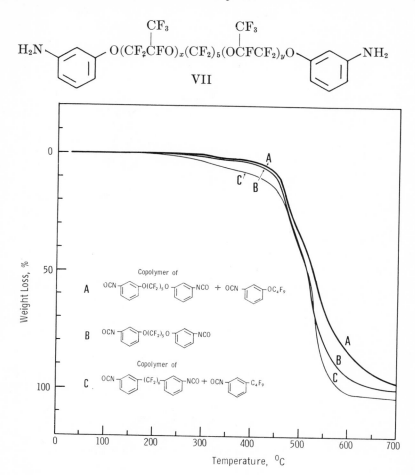

Figure 1.   Thermogravimetric analysis of isocyanurate-linked fluoro-carbon polymers, heating rate 2.7°C/minute in air.

Increasing chain length of the fluorocarbon ether segment was accompanied by increasing flexibility and lowering of the glass-transition temperature. As the chain segment length was increased from $x+y=0$ to $x+y=3$, the transition temperature decreased from about 125 to about 80°C.

High thermal and oxidative stability of the imide-linked fluoro-carbon ether polymers is shown by thermogravimetric analysis in air (Figure 2). The weight loss was less than 5% to a temperature of 450°C. Isothermal weight loss in air at 260°C was less than 0.3% in 112 hours for the polymer where $x+y=3$.

Tensile strength and elongation data for the polymer where $x+y=2$ and 3 are shown in Table I. These determinations were carried out

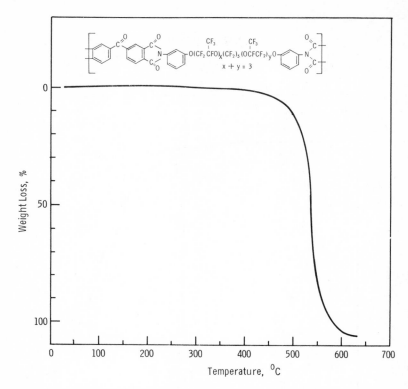

Figure 2. Thermogravimetric analysis of polyimide-linked perfluoro-alkylene ether polymer, heating rate 2.7°C/minute in air

Table I. Tensile Strength of Fluorocarbon Polyimide

Polymer[a]	Tensile test temp. (°C)	Tensile (psi)	Elongation (%)
$x + y = 2$ ($T_g = {\sim}100°$)	25	7500	160–200
$x + y = 3$ ($T_g = {\sim}80°$)	25	8000	280
	100	2700	${\sim}750$
	200	100	500
	25[b]	6800	240
	25[c]	5200	200

[a] From benzophenonetetracarboxylic dianhydride and

$$H_2N \text{—} \bigcirc \text{—} O(CF_2CFCF_3O)_x(CF_2)_5(OCFCF_3CF_2)_yO \text{—} \bigcirc \text{—} NH_2$$

[b] Sample heated for 70 hours in air at 260°C
[c] Sample refluxed in water for 70 hours

with microtensile test specimens that permit a reasonable approximation of results that can be expected with standard ASTM tensile specimens. The results in Table I indicate good strength and ability to withstand thermal and oxidizing conditions at 260° and hydrolysis at 100°C.

Experimental

Perfluoroalkylene telomer diiodide intermediates were purchased from the Thiokol Corp. Perfluoroalkylene ether dicarboxylic acid intermediates were purchased as a mixture of the acid fluorides from PCR, Inc. under an agreement with Dupont. Properties of distilled fractions are indicated in Table II. The discrepancy in observed and calculated equivalent weights results from the presence of close boiling mono acid fluoride by-products, $C_3F_7O(CFCF_2O)_nCFCF_3COF$.

$$\overset{|}{CF_3}$$

Hydrolysis of each acid fluoride fraction and redistillation of the carboxylic acids allowed separation of mono- from dicarboxylic acids (Table III). Thus, the 1:1 dicarboxylic acid and the monocarboxylic acid formed from three molecules of hexafluoropropylene epoxide were isolated from the 1:1 oligomer acid fluoride fraction shown in Table II. The acids were converted to acid chlorides in high yields (about 90%) by prolonged refluxing with thionyl chloride (Table IV). A trace of pyridine was added as a catalyst.

Nitrophenyl esters were prepared in high yield (about 90%) by reaction of excess (10% or so) m-nitrophenol with the respective acid chlorides in the presence of a trace of pyridine at 145°C. The physical properties are listed in Table V.

Fluorination of nitrophenyl esters was carried out in stainless steel and Hastelloy autoclaves. Sulfur tetrafluoride used in fluorination was treated with mercury (16) to remove bromine, a consistent contaminant of the commercially available gas. Purification was essential for high yields. The presence of sufficient anhydrous hydrogen fluoride to ensure the presence of a liquid phase also appears to be important in obtaining reasonably good yields. Fluorination experiments are summarized in Table VI.

In a typical experiment, a 1400-ml Hastelloy reactor was charged with m-nitrophenyl ester of 2:1 oligomer acid (50 grams, 0.0614 mole) and 500 ml of perfluorinated fluid FC-75 (3M Co.) as a solvent. The reactor was sealed, evacuated, and cooled in solid carbon dioxide. Purified sulfur tetrafluoride (50 grams, 0.46 mole) (toxic) and hydrogen fluoride (75 grams, 3.75 moles) were introduced. The reactor was rocked and heated for 66 hours at 80°C. After release of gases, the contents of the reactor were poured over ice and neutralized with $NaHCO_3$. Ether was added to extract and separate products. The FC-75 solvent separated cleanly as a third phase. The ether phase was separated and washed. The ether solution yielded 51.6 grams of crude product. This was treated further with aqueous 0.5N NaOH in the presence of ether. The alkali-treated product (46 grams) was soluble in benzene and showed no carbonyl band in the IR spectrum. Distilla-

Table II.   Perfluoroalkylene

Compound

$$\overset{\displaystyle CF_3}{\underset{\displaystyle |}{FOCCFO(CF_2)_4COF}}{}^{a}$$ (1:1 oligomer)

$$\overset{\displaystyle CF_3 \qquad\quad CF_3}{\underset{\displaystyle | \qquad\qquad |}{FOCCFO(CF_2)_5OCFCOF}}{}^{b}$$ (2:1 oligomer)

$$\overset{\displaystyle CF_3 \qquad\quad CF_3 \qquad CF_3}{\underset{\displaystyle | \qquad\qquad | \qquad\quad |}{FOCCFO(CF_2)_5OCFCF_2OCFCOF}}$$ (3:1 oligomer)

^{a} Mixture of mono- and dicarboxylic acids

Table III.   Perfluoroalkylene

Compound	Percent yield
$\overset{CF_3}{\underset{|}{HOOCCFO(CF_2)_4COOH}}$	47
$\overset{CF_3 \quad CF_3}{\underset{| \qquad |}{HOOCCFOCF_2CFOC_3F_7}}$	30
$\overset{CF_3 \qquad CF_3}{\underset{| \qquad\quad |}{HOOCCFO(CF_2)_5OCFCOOH}}$	87
$\overset{CF_3 \qquad CF_3 \quad CF_3}{\underset{| \qquad\quad | \qquad |}{HOOCCFO(CF_2)_5OCFCF_2OCFCOOH}}$	92

Table IV.   Fluorocarbon Ether Acid Chlorides

Compound	Percent yield	bp °C/torr	n_D^{25}
$\overset{CF_3}{\underset{|}{ClOCCFO(CF_2)_4COCl}}$	68	157/740	1.3319
$\overset{CF_3 \quad CF_3}{\underset{| \qquad |}{ClOCCFO(CF_2)_5OCFCOCl}}$	87	87/14	1.3320
$\overset{CF_3 \quad CF_3 \quad CF_3}{\underset{| \qquad | \qquad |}{ClOCCFO(CF_2)_5OCFCF_2OCFCOCl}}$	87	117/17	1.3176

Ether Dicarboxylic Acid

bp °C/torr	$n_D{}^{25}$	Neutral equivalent	
		Calc.	Found
108–109/740	<1.3	102.5	150[a]
154/740 65–67/24	<1.3	144.0	160
96–98/24	<1.3	185.4	188

[b] Mixture of symmetrical and unsymmetrical oligomer

Fluoride Intermediates

bp °C/torr	$n_D{}^{25}$	Neutral equiv.	
		Calc.	Found
93–95/0.02	1.3372	203	208
85/12	1.3058	496	506
126/0.04	1.3229	286	280
140/0.25	1.3186	369	366

tion gave 13 grams of an impure fraction and 25 grams of nitrophenoxy product boiling ~180°C/0.06 mm, $n_D{}^{25}$ 1.4098, 47% yield. Redistillation of the impure fraction gave an additional 9 grams of material, $n_D{}^{25}$ 1.4100, 65% yield.

Catalytic hydrogenation of nitrophenyl derivatives over Raney nickel at 40 psig formed the corresponding amines in near quantitative yields. Properties of distilled products are shown in Table VII. Aromatic amine equivalents were determined in glacial acetic acid by titration using standard $HClO_4$ in glacial acetic acid titrant with crystal violet indicator. Phosgenation of perfluoroalkylenearylamine hydrochlorides in refluxing xylene or dichlorobenzene formed diisocyanate deriva-

Table V. Nitrophenyl

Compound

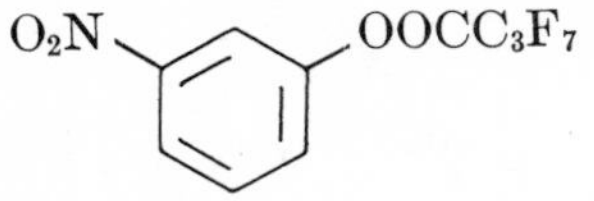

O_2N———$OOCC_3F_7$

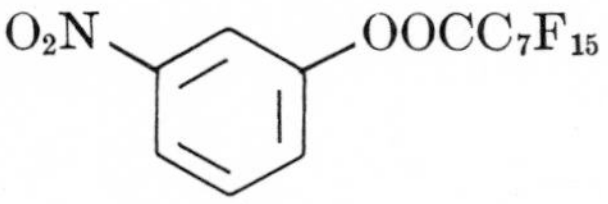

O_2N———$OOCC_7F_{15}$

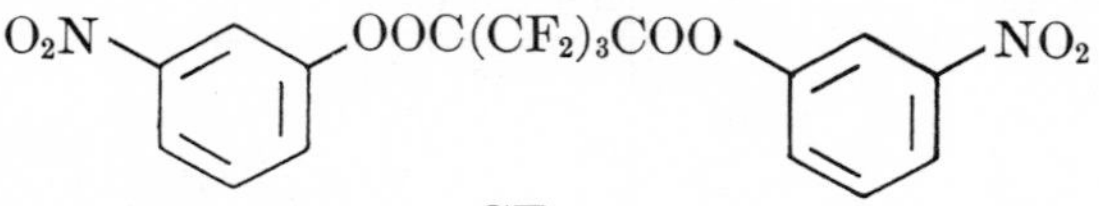

O_2N———$OOC(CF_2)_3COO$———NO_2

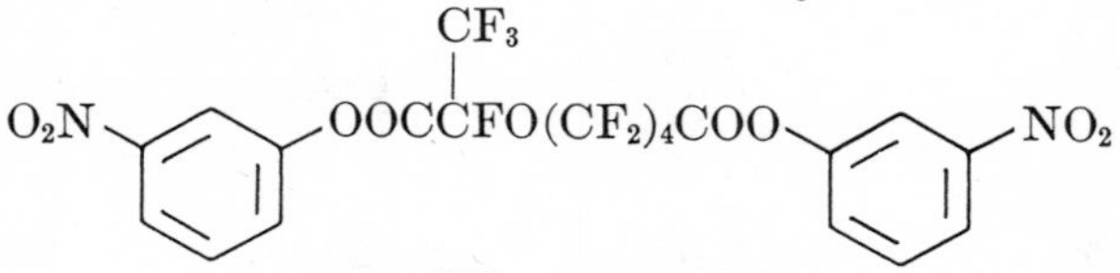

$\overset{CF_3}{|}$
O_2N———$OOCCFO(CF_2)_4COO$———NO_2

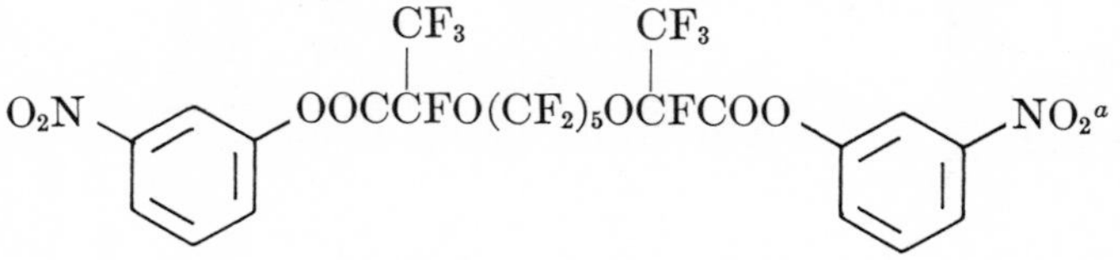

$\overset{CF_3}{|} \qquad \overset{CF_3}{|}$
O_2N———$OOCCFO(CF_2)_5OCFCOO$———$NO_2{}^a$

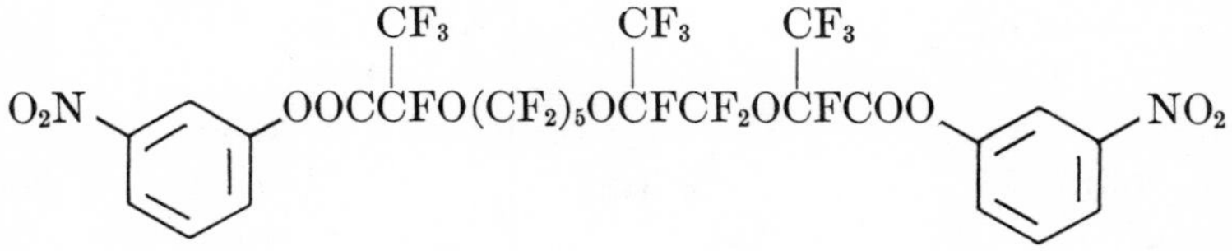

$\overset{CF_3}{|} \qquad \overset{CF_3}{|} \qquad \overset{CF_3}{|}$
O_2N———$OOCCFO(CF_2)_5OCFCF_2OCFCOO$———$NO_2$

[a] Anal: Calcd. for $C_{23}H_8F_{18}N_2O_{10}$: C, 33.92; H, 0.99; N, 3.44
Found: C, 34.57; H, 1.06; N, 3.34

Table VI. Nitrophenyl

Nitrophenyl Intermediates

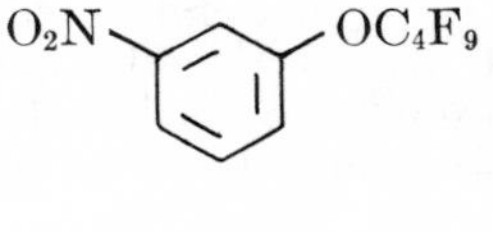

O_2N———OC_4F_9

Ester Intermediates

% Yield	bp °C/torr	mp °C	n_D^{25}
89	63/0.14		1.4281
92	95–97/0.08	36–38.5	
85	197–205/0.04	97–100	
40	190–200/0.03		1.4578
90	195/0.05	60–80	
92	200/0.05	∼35	1.4080

Intermediates

Percent yield	bp °C/torr	mp °C	n_D^{25}
62	67/0.2		1.4027

Table VI.

Nitrophenyl Intermediates

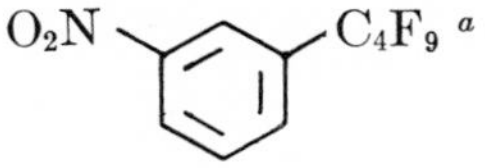

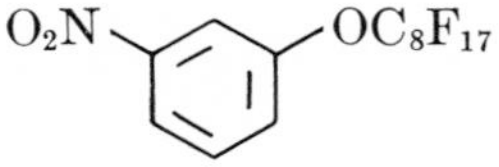

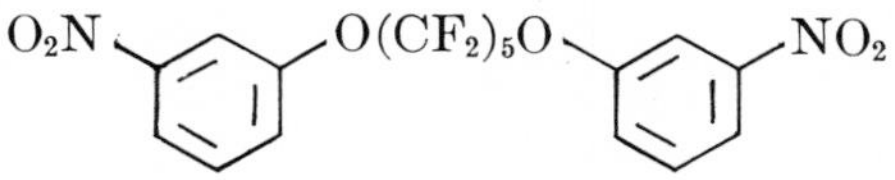

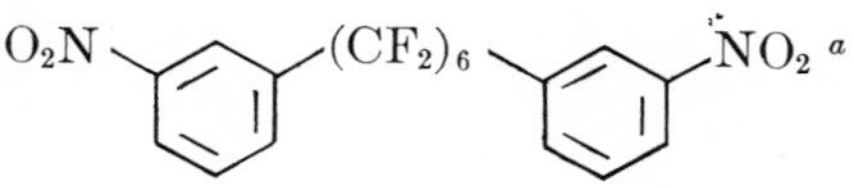

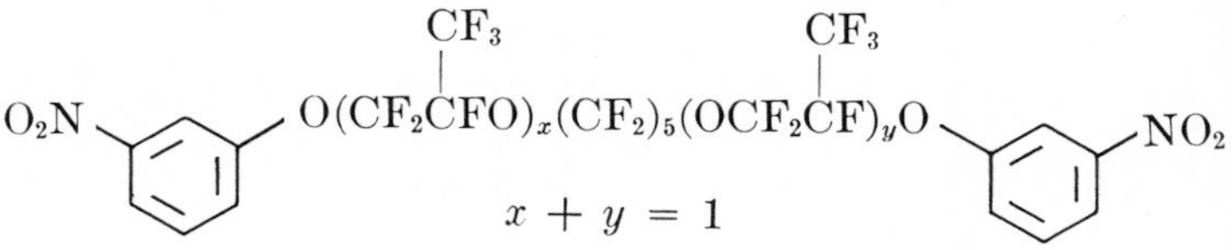

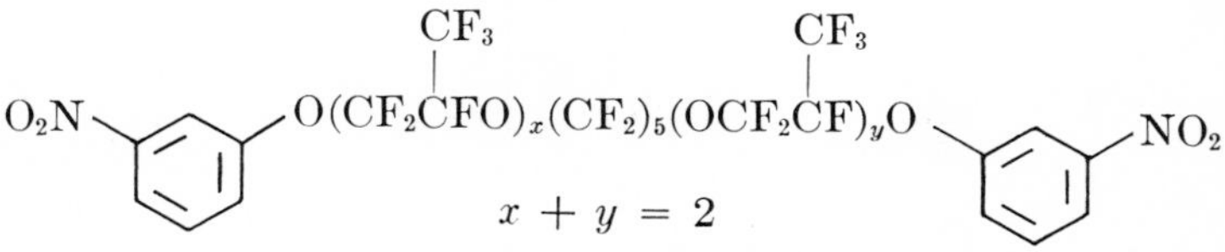

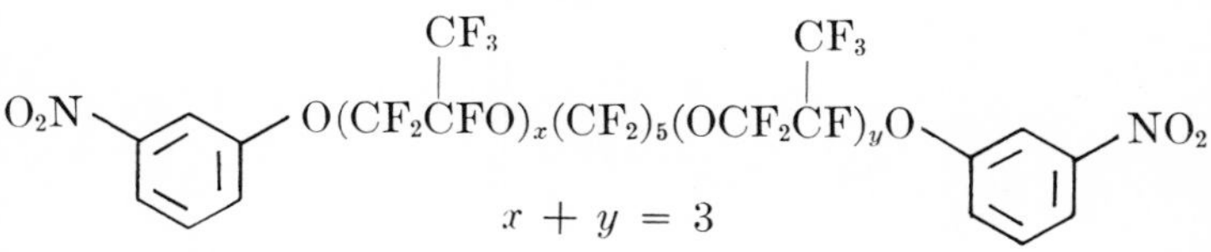

O$_2$N—⟨C$_6$H$_4$⟩—C$_4$F$_9$ [a]

O$_2$N—⟨C$_6$H$_4$⟩—OC$_8$F$_{17}$

O$_2$N—⟨C$_6$H$_4$⟩—O(CF$_2$)$_5$O—⟨C$_6$H$_4$⟩—NO$_2$

O$_2$N—⟨C$_6$H$_4$⟩—(CF$_2$)$_6$—⟨C$_6$H$_4$⟩—NO$_2$ [a]

$$\text{O}_2\text{N}-\langle\text{C}_6\text{H}_4\rangle-\text{O(CF}_2\overset{\overset{\text{CF}_3}{|}}{\text{C}}\text{FO)}_x\text{(CF}_2)_5\text{(OCF}_2\overset{\overset{\text{CF}_3}{|}}{\text{C}}\text{F)}_y\text{O}-\langle\text{C}_6\text{H}_4\rangle-\text{NO}_2$$
$$x + y = 1$$

$$\text{O}_2\text{N}-\langle\text{C}_6\text{H}_4\rangle-\text{O(CF}_2\overset{\overset{\text{CF}_3}{|}}{\text{C}}\text{FO)}_x\text{(CF}_2)_5\text{(OCF}_2\overset{\overset{\text{CF}_3}{|}}{\text{C}}\text{F)}_y\text{O}-\langle\text{C}_6\text{H}_4\rangle-\text{NO}_2$$
$$x + y = 2$$

$$\text{O}_2\text{N}-\langle\text{C}_6\text{H}_4\rangle-\text{O(CF}_2\overset{\overset{\text{CF}_3}{|}}{\text{C}}\text{FO)}_x\text{(CF}_2)_5\text{(OCF}_2\overset{\overset{\text{CF}_3}{|}}{\text{C}}\text{F)}_y\text{O}-\langle\text{C}_6\text{H}_4\rangle-\text{NO}_2$$
$$x + y = 3$$

[a] *Via* nitration of the perfluoroaromatic compound

tives in high yields (Table VIII). Isocyanate equivalents were determined by reaction of the isocyanate compound with a known excess of dry dibutylamine in toluene solution, followed by titration with standard HCl.

Continued

Percent Yield	bp °C/torr	mp °C	n_D^{25}
95	113/17		1.4115
69	103/0.1		1.3804
87	190/0.05		1.4729
85		100–102	
65	205–210/0.015		1.4355
65	177/0.06		1.4095–1.4110
50	179/0.03		1.3944

Polymerizations. Isocyanurate polymer formation was catalyzed by the addition of N,N,N',N'-tetramethyl-1,3-butanediamine and allyl glycidyl ether (redistilled); 1% (by weight) of each catalyst was used. Polymerization occurred at 25° to 50°C under anhydrous conditions although temperatures of 100° to 150°C were needed to complete the reaction with the short chain diisocyanate intermediates used.

Table VII. Fluorocarbon and

Amine compound

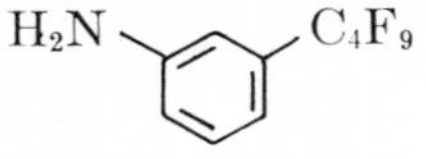

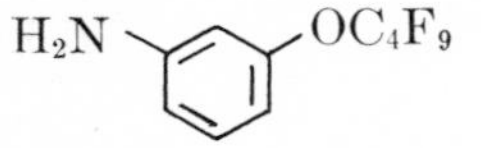

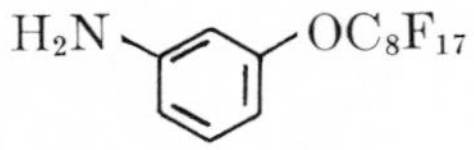

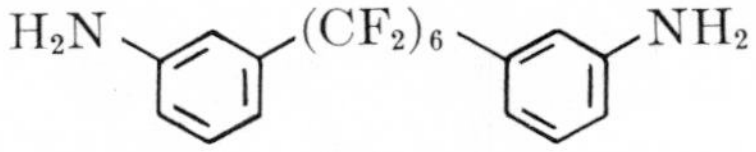

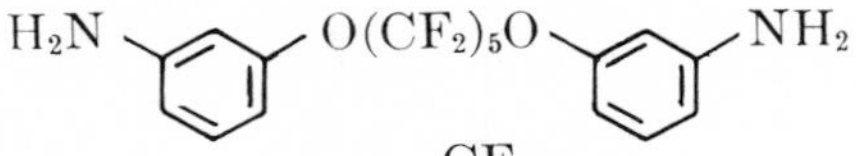

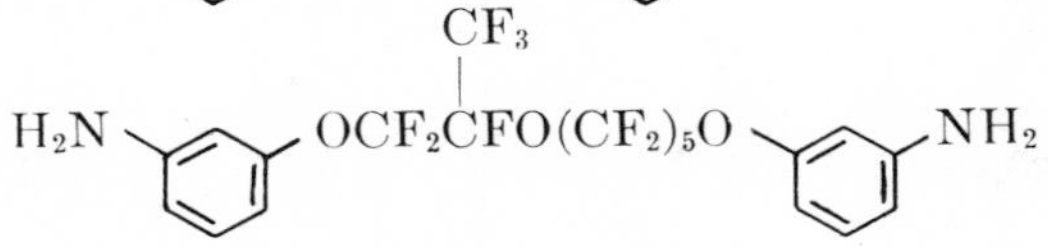

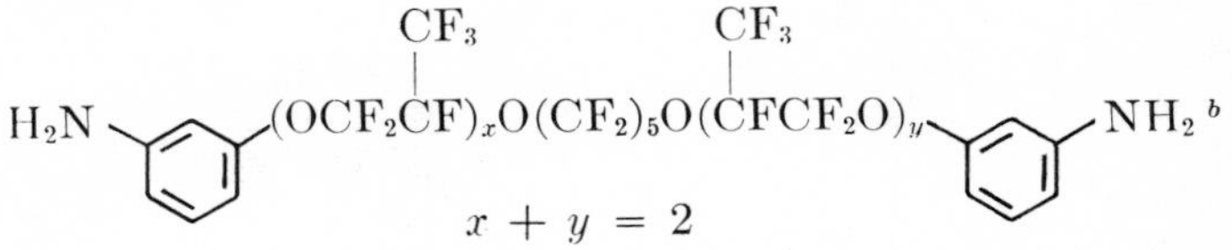

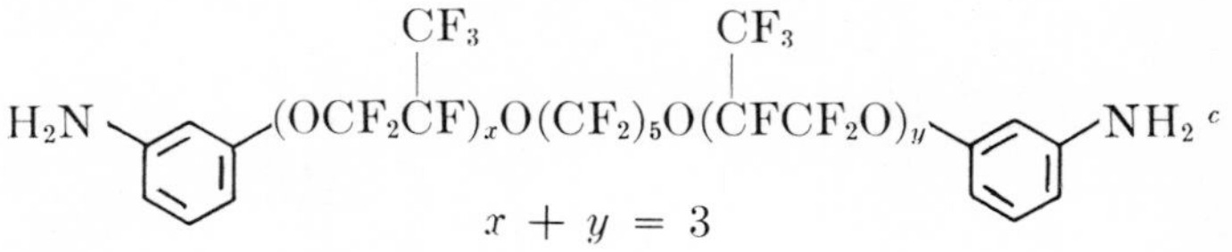

H_2N—〈 〉—C_4F_9

H_2N—〈 〉—OC_4F_9

H_2N—〈 〉—OC_8F_{17}

H_2N—〈 〉—$(CF_2)_6$—〈 〉—NH_2

H_2N—〈 〉—$O(CF_2)_5O$—〈 〉—NH_2

$$\overset{\displaystyle CF_3}{|}$$
H_2N—〈 〉—$OCF_2CFO(CF_2)_5O$—〈 〉—NH_2

$$\overset{\displaystyle CF_3}{|} \qquad\qquad \overset{\displaystyle CF_3}{|}$$
H_2N—〈 〉—$(OCF_2CF)_xO(CF_2)_5O(CFCF_2O)_y$—〈 〉—$NH_2$ [b]

$$x + y = 2$$

$$\overset{\displaystyle CF_3}{|} \qquad\qquad \overset{\displaystyle CF_3}{|}$$
H_2N—〈 〉—$(OCF_2CF)_xO(CF_2)_5O(CFCF_2O)_y$—〈 〉—$NH_2$ [c]

$$x + y = 3$$

[a] Crude product after removing solvent
[b] Anal: Calc. for $C_{23}H_{12}F_{22}N_2O_4$: C, 34.60; H, 1.52; F, 52.36; N, 3.51
                  Found: C, 34.68; H, 1.43; F, 52.43; N, 3.41

Table VIII. Isocyanate

Isocyanate

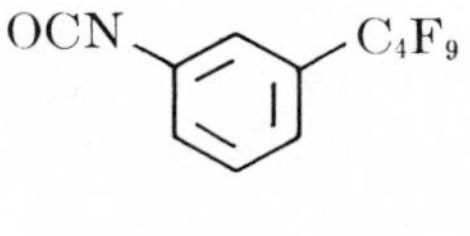

OCN—〈 〉—C_4F_9

Fluorocarbon Ether Arylamines

Percent yield	bp °C/torr	mp °C	n_D^{25}	Equiv. wt. Found	Equiv. wt. Calc.	Amine·HCl mp °C
91			1.4107[a]			150–152
97			1.3998[a]			215–220
85		188–196[a]				195–217
93		82–86[a]				261–268
95	162/0.03	47–47.5	1.4755	236	233	264
72			1.4368[a]	331	316	
95	157/0.007		1.4082	410	399	
89	166/0.008		1.3920	490	482	

[c] Anal: Calc. for $C_{26}H_{12}F_{28}N_2O_5$: C, 32.38; H, 1.25; F, 55.17; N, 2.91
Found: C, 32.45; H, 1.08; F, 55.53; N, 3.20

Intermediates

Percent yield	bp °C/torr	mp °C	n_D^{25}	Isocyanate equivalent Found	Isocyanate equivalent Calc.
73	145/133		1.4111	343	337

Table VIII.

Isocyanate

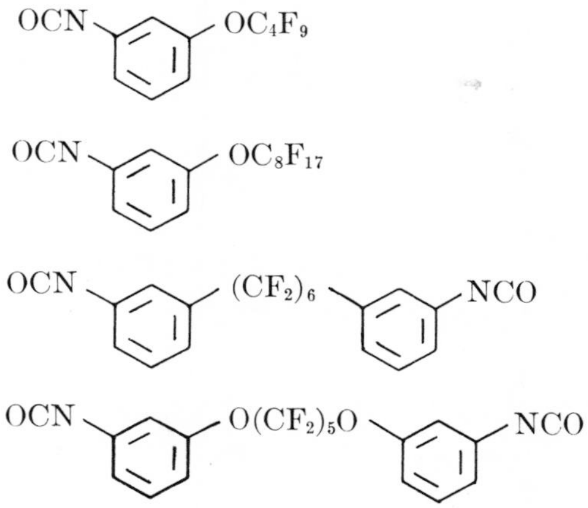

The polyimide polymers were formed by an established procedure (17) involving the addition of stoichiometric quantities of benzophenone tetracarboxylic dianhydride (sublimed) to the diamine dissolved in anhydrous dimethylacetamide. The polyamic acid solution was cas on Teflon, warmed (80° to 90°C) to drive off solvent, then gradually heated to a temperature of 180°C. The polyimide forms rapidly at 120 to 150°C.

Analysis of polymer, $x + y = 3$.

Calc. for $C_{43}H_{14}F_{28}N_2O_{10}$:   C, 41.30; H, 1.13; F, 42.54; N, 2.24
Found: C, 41.17; H, 1.07; F, 42.15; N, 2.11

Acknowledgment

This work was supported by the George C. Marshall Space Flight Center of the National Aeronautics and Space Administration under Contract NAS8-21401. The authors are pleased to acknowledge the contribution of William J. Patterson, MSFC Contracting Officer's Technical Representative, in terms of his interest, suggestions, and stimulating discussions.

Literature Cited

1. Sheppard, W. A., *J. Org. Chem.* (1964) **29**, 1.
2. Dorfman, E., Emerson, W. E., Carr, R. L. K., Bean, C. T., *Rubber Chem Tech.* (1966) **39**, 1175.
3. Griffin, W. R., *Rubber Chem. Tech.* (1966) **39**, 1178.

ontinued

				Isocyanate Equivalent	
				---	---
Percent Yield	*bp °C/torr*	*mp °C*	n_D^{25}	*Found*	*Calc.*
50	152/150		1.4009		
58	77/0.05		1.3790–1.3799		
86	152–153/0.03	72.5–74.5		268	268
83	153/0.03		1.4690	261	259

4. Graham, T. L., *Rubber Age* (1969) **101(8)**, 43.
5. Kim, Y. K., Pierce, O. R., Bajzer, W. X., Smith, A. G., *Polymer Preprints* (1971) **12(1)**, 482.
6. Zollinger, J. L., Throckmorton, J. R., Ting, S. T., Mitsch, R. A., Elrick, D. E., *J. Macromol. Sci. Chem.* (1969) **A-3**, 1443.
7. Critchley, J. P., McLoughlin, V. C. R., Thrower, J., White, I. M., *Chem. Ind.* (1969) 934.
8. Yakubovich, A. Ya., Gitina, R. M., Zaitseva, Ye. L., Markova, G. S., Simonov, A. P., *Vysokomol. Soedin* (1970) **A12**, 2520.
9. Feast, W. J., Musgrave, W. K. R., Reeves, N., *J. Polym. Sci., Pt. A-1* (1971) **9**, 2733.
0. Fritz, C. G., Warnell, J. L., U.S. Patent **3,317,484** (1967).
1. Toy, M. S., *Polymer Preprints* (1971) **12(1)**, 385.
2. McLoughlin, V. C. R., Thrower, J., *Tetrahedron* (1969) **25**, 5921.
3. Zaitseva, E. L., Yakubovich, A. Ya., *J. Gen. Chem. USSR* (1966) **36**, 359.
4. Gilman, L. G., Gollis, M. H., U.S. Patent **3,211,704** (Oct. 1965).
5. Blake, E. S., Hammann, W. C., Edwards, J. W., Reichard, T. E., Ort, M. R., *J. Chem. Eng. Data* (1961) **6**, 87.
6. Arth, G. E., Fried, J., U.S. Patent **3,046,094** (1962).
7. Sorenson, W. R., Campbell, T. W., "Preparative Methods of Polymer Chemistry," Interscience, New York, 1968.

ECEIVED March 29, 1972.

6

Properties of Polyphenylene Sulfide Coatings

H. WAYNE HILL, JR., and J. T. EDMONDS, JR.

Phillips Petroleum Co., Bartlesville, Okla. 74004

Polyphenylene sulfide is a unique material that combines high thermal stability with outstanding chemical resistance. This combination of properties provides unusual utility as molding resins and as protective coatings for the chemical and petroleum industries, and as release coatings in the food industry. The synthesis of phenylene sulfide polymers is presented here. Coatings of polyphenylene sulfide may be applied to a variety of substrate metals by slurry spraying, fluidized-bed techniques, and dry-powder spraying. Release coatings based on polyphenylene sulfide may be obtained by adding a small amount of polytetrafluoroethylene to the coating formulation. The performance of these release coatings on household cookware is discussed.

$\mathbf{P}$olyphenylene sulfide is an unusual material, combining some of the characteristics of both thermoplastics and thermosets with an outstanding balance of high-temperature performance and chemical resistance. Although the material is an excellent molding resin, there are equally important applications as protective coatings for the chemical and petroleum industries and as nonstick coatings for the food and cookware industry.

In early work on organosulfur compounds, Duess (1) and Hilditch (2) reported the preparation of various aromatic disulfides by condensation reactions of thiophenol on treatment with aluminum chloride and sulfuric acid, respectively. Macallum (3) was the first to report the preparation of a phenylene sulfide polymer. His procedure involved the reaction of sulfur, sodium carbonate, and dichlorobenzene in a sealed vessel. Polymers made by this scheme generally have more than one sulfur atom between benzene rings, as indicated by the structure $-(C_6H_4S_x)_n-$.

Lenz and coworkers (4-6) have described the preparation of polyphenylene sulfide by a nucleophilic substitution reaction involving self condensation of materials such as copper *p*-bromothiophenoxide.   These substitution reactions are carried out at 200-250°C under nitrogen in the solid state, or in the presence of materials such as pyridine as reaction media.

The above two methods as well as other methods of polymerization have been reviewed by Smith (7).   We have discovered a new process for the preparation of a wide variety of polyarylene sulfides (8).   For example, polyphenylene sulfide may be prepared by the reaction of *p*-dichlorobenzene and sodium sulfide in a polar solvent.

$$Cl-\left\langle\right\rangle-Cl + Na_2S \xrightarrow[\text{Solvent}]{\text{Heat}} \left(\left\langle\right\rangle-S\right)_n + 2NaCl$$

Properties

In the Phillips process, polyphenylene sulfide (PPS) is obtained from the polymerization mixture in the form of a fine white powder, which, after purification, is designated Ryton V PPS.   Characterization of this polymer is complicated by its extreme insolubility in most solvents.   At elevated temperatures, however, Ryton V PPS is soluble to a limited extent in some aromatic and chlorinated aromatic solvents and in certain heterocyclic compounds.   The inherent viscosity, measured at 206°C in 1-chloronaphthalene, is generally 0.16, indicating only moderate molecular weight.   The polymer is highly crystalline, as shown by x-ray diffraction studies (9).   The crystalline melting point determined by differential thermal analysis is about 285°C.

When the molten polymer is subjected to additional heat in the presence of air, the melt darkens and, after a while, it gels and solidifies. The solid polymer is believed to be crosslinked because it is insoluble in all organic solvents, even at elevated temperature.

The changes that occur at high temperatures can be demonstrated readily by differential thermal analysis (DTA).   When the DTA is conducted under nitrogen to suppress crosslinking, the sample can be melted, cooled, and remelted with little effect on the thermal transitions. This experiment is illustrated by curve A of Figure 1, in which both the glass-transition temperature at 80°C and the crystalline melting point at 282°C are visible.   When the DTA is run on a sample that has been heated in air at 370°C for four hours, the crystalline melting point is barely visible in the DTA trace.   The sharp exotherms at 125°-135°C are indicative of crystallization temperatures.

While the linear polyphenylene sulfide polymer possesses a moderate degree of mechanical strength as it is produced in the polymerization process, it can be converted into a much tougher product by thermal treatment. Accordingly, when the polymer is heated to a high enough temperature in air, chain extension and crosslinking occur, producing a "cured" polymer that is tough, ductile, and very insoluble.

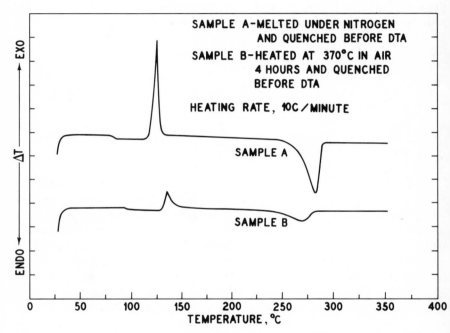

Figure 1. Differential thermal analysis of polyphenylene sulfide in nitrogen

The chemistry of this curing process involves several complex reactions. In addition, the very limited solubility of the linear polymer and the extreme insolubility of the cured polymer make exact structural assignments almost impossible. It is possible to describe some of the contributing reactions in qualitative terms. For example, a chain-extension reaction involving thermal scission of carbon-sulfur bonds near the end of a polymer chain, followed by formation of a new carbon-sulfur bond between two large polymer residues and between two small polymer residues, can lead to an overall increase in molecular weight when the small molecules formed are lost by vaporization. This process is essentially an exchange reaction, as indicated in this structure, where m is significantly larger than n:

$$2-\left(\!\!\left\langle \underline{} \right\rangle\!\!-S\right)_m-\left(\!\!\left\langle \underline{} \right\rangle\!\!-S\right)_n \rightarrow \left(\!\!\left\langle \underline{} \right\rangle\!\!-S\right)_{2m} + \left(\!\!\left\langle \underline{} \right\rangle\!\!-S\right)_{2n} \uparrow$$

I   II

Thus, II is lost at the high temperatures involved in the curing process. Other possible reactions also occur. For example, oxidative coupling between aromatic rings (biphenyl reaction); nucleophilic attack on an aromatic ring of one polymer chain by an end-group function of another polymer chain, or by a cleaved segment derived from another polymer chain; sulfonium ion formation involving a sulfide link and a sulfur-containing end group. Reactions of this type produce an increase in molecular weight and crosslinking. It is likely that several of these reactions occur simultaneously during the curing process.

Thermogravimetric analysis of polyphenylene sulfide in nitrogen or in air indicates no appreciable weight loss below about 500°C. Degradation is essentially complete in air at 700°C, but in an inert atmosphere, about 40% of the polymer weight remains at 1000°. In addition, polyphenylene sulfide prepared by the Phillips process is more stable than that prepared by the Lenz process (6). Comparative thermogravimetric data, shown in Figure 2, demonstrate that polyphenylene sulfide displays a much greater resistance to weight loss at elevated temperatures than do either the conventional thermoplastics or such specialty heat-resistant polymers as polytetrafluoroethylene. Mechanical properties of molded specimens are essentially unaffected after exposure at 450°F in air for four months.

Polyphenylene sulfide also possesses unusual chemical resistance. To demonstrate this resistance, injection-molded tensile bars of cured polymer were exposed to a representative groups of reagents at 200°F for 24 hours. After exposure, the bars were weighed to determine weight gain or loss, and the tensile strength determined. The results of these experiments are given in Table I.

The tensile strength of the unexposed bars averaged 11,000 psi. These tensile bars are unaffected by gasoline, motor oil, hydrocarbons, and carbon tetrachloride, while exposure to trichloroethylene results in a small weight gain and loss in tensile strength. Alcohols, ketones, esters, and organic acids do not affect the polymer. Some nitrogenous organic compounds—such as butylamine, pyridine, and acetonitrile—cause a modest loss in tensile strength. Other chemicals, such as dimethylaniline, ethanolamine, and nitrobenzene, have virtually no effect on tensile strength or weight change. In general, oxidizing agents such as bromine water, aqua regia, and 50% chromic acid cause a severe loss

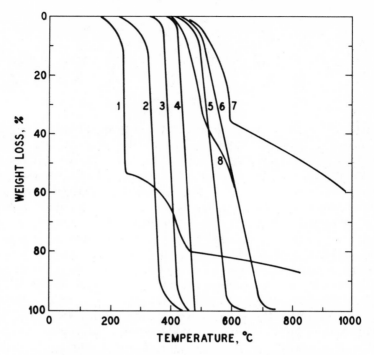

KEY:1-POLYVINYLCHLORIDE;2-POLYMETHYLMETHACRYLATE;
3-POLYSTYRENE; 4-POLYETHYLENE;5-POLYTETRAFLUORO-
ETHYLENE; 6-POLYPHENYLENESULFIDE IN AIR ATMOSPHERE;
7-POLYPHENYLENE SULFIDE; 8-POLYPHENYLENE SULFIDE
(LENZ⁴ᶜ) IN AIR.

*Figure 2.   Comparative thermogravimetric analysis (nitrogen
atmosphere)*

in strength.   However, aqueous potassium dichromate does not.

Strong acids, such as 96% sulfuric acid, attack the polymer rather severely; weaker acids and a variety of inorganic bases are not detrimental.   The polymer is inert to a wide variety of inorganic salt solutions, with some samples of coated metals having been exposed in salt water in the Gulf of Mexico for one year without deterioration. In fact, even the uncured polymer is remarkably resistant to a broad group of chemicals.

Applications

Injection-Molding Resins. A high melt viscosity resin is recommended for use in injection-molding applications.   The mechanical

Table I.   Chemical Resistance of PPS Tensile Bars

Exposure: 200°F for 24 hours

Test Chemical	Weight Change (%)	Tensile [a] (psi)
Hydrocarbons and Chlorinated Hydrocarbons		
Kerosene	−0.05	11,700
Motor Oil	−0.02	11,900
Carbon Tetrachloride	1.7	11,100
Cyclohexane	0.05	12,000
Gasoline	0.07	10,300
Trichloroethylene	6.52	7,400
Alcohols, Ketones, Esters, and Ethers		
Butyl Alcohol	0.05	10,500
Methyl Ethyl Ketone	1.02	11,200
Amyl Acetate	0.14	13,300
Dioctyl Phthalate	−0.07	13,100
Dibutyl Ether	0	11,400
Organic Acids		
Glacial Acetic	0.09	12,400
Trichloroacetic	0.45	12,000
Formic (88%)	0.18	10,900
Benzenesulfonic	−0.05	11,400
Nitrogenous Organic Compounds		
Butylamine	1.52	7,100
Dimethylaniline	2.32	10,200
Ethanolamine	0	12,600
Pyridine	3.84	8,900
Acetonitrile	0.59	9,000
Nitrobenzene	2.43	11,000
Oxidizing Agents		
Bromine Water	4.34	Severe attack
Aqua Regia (Room Temperature Exposure)	18.64	Severe attack
50% Chromic Acid	0.84	3,300
10% Potassium Dichromate	0.36	12,300
Sodium Hypochlorite (Clorox)	0.50	5,400
Inorganic Acids and Bases		
10% Nitric Acid	0.32	12,000
37% Hydrochloric Acid	0.57	10,900
30% Sulfuric Acid	0.14	11,700
96% Sulfuric Acid	Severe attack	Severe attack
85% Phosphoric Acid	−0.05	12,500
10% Sodium Bicarbonate	0.36	11,600

Table I.　Continued

Exposure: 200°F for 24 hours

Test Chemical	Weight Change (%)	Tensile [a] (psi)
10% Sodium Carbonate	0.32	9,600
30% Sodium Hydroxide	0.07	10,000
78% Ammonium Hydroxide	0.73	11,400
Inorganic Salt Solutions		
Saturated Sodium Chloride	0.16	10,400
10% Sodium Acetate	0.32	13,800
10% Sodium Nitrate	0.32	11,100
10% Sodium Sulfate	0.36	10,100
Trisodium Phosphate	0.36	11,200
10% Calcium Chloride	0.36	12,700

[a] Tensile of unexposed specimen was 11,000 psi

properties of unfilled and glass-filled injection-molded specimens are summarized in Table II. The unfilled resin is characterized by high tensile strength, high flexural modulus, high heat-deflection temperature, and modest impact strength. The glass-filled resin has superior properties and is generally finding greater practical applications than is the unfilled material. For example, the 40% glass-filled resin has a tensile strength of 21,000 psi at room temperature, and 4700 psi at 400°F. Its tensile strength at 400°F is greater than that of polyethylene at room temperature. The flexural modulus is 2.2×10^6 psi at room temperature, decreasing gradually to 600,000 psi at 450°F. The flexural modulus at 450° F is greater than the room-temperature flexural modulus of many established plastics, such as ABS resins, polyacetals, nylons, and polycarbonates.

As a further comparison, the flexural modulus of glass-filled polyphenylene sulfide at 450°F is about 10 times that of unfilled polytetrafluoroethylene at room temperature. These data illustrate the outstanding retention of stiffness of this material at elevated temperatures. The heat-deflection temperature of polyphenylene sulfide containing 40% glass fibers is greater than 425°F, accounting for the excellent retention of mechanical properties at elevated temperatures.

Electrical properties of polyphenylene sulfide compounds are summarized in Table III. The dielectric constant of 3.1 is low in comparison with many other plastic materials. Similarly, the dissipation factor is very low. Dielectric strength ranges from about 500-600 volts per mil for the various compounds; these values are quite high. Thus, both filled and unfilled polyphenylene sulfide materials are excellent electrical insulators.

Limiting oxygen index values (LOI) of a number of plastics are shown in Table IV. The LOI is the concentration of oxygen required to maintain burning. Polyphenylene sulfide has a value of 44, and falls among the least flammable types of plastics.

Table II. Mechanical Properties of Injection-Molding Compositions

	Ryton PPS	Ryton PPS Glass-Filled (60/40)
Density	1.34	1.64
Tensile, psi		
At 70°F	11,000	21,000
At 400°F —	4,700	4,700
Elongation (70°F), %	3	3
Flexural Modulus, psi		
At 70°F	600,000	2,200,000
At 450°F		600,000
Flexural Strength, psi	20,000	37,000
Hardness, Shore D	86	92
Notched Izod Impact, ft lb/in		
At 75°F	0.3	0.8
At 300°F	1.0	1.8
Heat Deflection Temperature @ 264 psi, °F	280	>425
Maximum Recommended Service Temperature, °F	500	500

Table III. Electrical Properties of Polyphenylene Sulfide Compounds

	Unfilled PPS	40% Glass-Filled PPS
Dielectric Constant		
10^3 Hertz	3.1	3.8
10^6 Hertz	3.1	3.8
Dissipation Factor		
10^3 Hertz	0.0004	0.0037
10^6 Hertz	0.0007	0.0066
Dielectric Strength, Volts/Mil	585	490

Table IV.   Flammability

Material	Limiting Oxygen Index, %
Polyvinyl chloride	47
Polyphenylene sulfide	44
Nylon 6–6	28.7
Polycarbonate	25
Polystyrene	18.3
Polyolefins	17.4
Polyacetal	16.2

Coatings. Polyphenylene sulfide coatings are characterized by an unusual combination of thermal stability and chemical resistance. They are finding acceptance as corrosion-resistant coatings for metals in the chemical and petroleum industries. Polyphenylene sulfide coatings may be applied by many different solventless coating systems such as slurry spraying, fluidized-bed coating, and flocking. All of these techniques require a bake cycle to cure the polymer to obtain tough, coalesced coatings. Typical bake cycles of 45 minutes at 700°F or 15 minutes at 800°F in a circulating air oven are usually enough to cure coatings of about 5-mil thickness or less. Thicker coatings require additional baking to achieve a sufficient level of curing to provide optimum mechanical toughness.

Best adhesion is obtained when the metal surface is grit-blasted before coating. Aluminum surfaces do not require any additional surface treatment for good adhesion. For adequate adhesion, iron surfaces should be heat-treated at 700°F in air before slurry coating, or primed with a mixture of polyphenylene sulfide and cobalt oxide in slurry form before application of a coating by fluidized-bed techniques. Uncured polymer should be used for application of thin coatings (1-2 mils per coat) by slurry techniques; the cured resins are preferred for application of thicker coatings by fluidized-bed or flocking techniques to avoid problems of dripping and sagging during the operation. Pigments such as titanium dioxide, channel black, and a variety of iron oxide compositions may be used when desired.

Two typical slurry coating formulations are shown in Table V. Formulation A, which consists of PPS, TiO_2, water, and a dispersing agent, is suitable for the production of multiple coats and for a variety of other applications. Formulation B contains polytetrafluoroethylene, and can be used to produce single coat nonstick surface coatings, or it can be applied as a top coat to a surface already coated with Formulation A.

Ryton PPS coatings that are pinhole free at 2-4 mils may be obtained easily by slurry-coating procedures. Properties of these coatings are indicated in Table VI. These coatings are quite hard, with a pencil hardness of 2H at room temperature. This hardness is maintained at temperatures up to 300°F, and the coating still has a pencil hardness of 2B at 500°F. Hardness and adhesion to grit-blasted aluminum, as measured on the Arco Microknife, are 450-600 grams and 4-5 mils, respectively, for a coating containing 3 parts PPS and 1 part TiO_2. These coatings will withstand forward and reverse impact tests of 160 inch-pounds, and a 3/16 inch mandrel bend of 180°. Elongation of the coating is greater than 32% (limit of test), as determined on a conical

Table V.   Polyphenylene Sulfide Formulations for Spray Coating Applications

	Parts by Weight	
	A	B
Ryton V PPS	100	100
Titanium Dioxide	33	33
Polytetrafluoroethylene	—	10
Water	300	300
Triton X-100	3	3
Cure Conditions		
Temperature, °F	700	700
Time, Minutes	45	45

Table VI.   Properties of Polyphenylene Sulfide Coatings

Property	PPS/TiO_2 Coating (3/1)	$PPS/TiO_2/$ PTFE Coating (3/1/0.3)
Hardness, Pencil	2H	2H
Hardness, Arco Microknife [a], g	500	350
Adhesion, Arco Microknife [b], mils	5	4–6
Mandrel Bend, 180°, 3/16″	Pass	Pass
Elongation (ASTM D 522), %	>32	>32
Reverse Impact, inch-pounds	160	160
Abrasion Resistance, Taber mg loss/1000 rev., CS–17 Wheel	50	57
Contact Angle [c], water, degrees	82	110
Contact Angle [c], Wesson oil, degrees	41	68
Chemical Resistance	Excellent	Excellent
Thermal Stability	Excellent	Excellent
Color	Light Tan	Light Tan

[a] ASTM 2197; measured on 1-mill coatings
[b] ASTM 2197; measured on aluminum coupons that had been grit-blasted before coating (coating thickness, 1 mil).
[c] Measured with a Rame-Hart Model A-100 goniometer

mandrel test apparatus. Thus hardness, toughness, and extensibility are excellent.

Excellent release coatings based on polyphenylene sulfide may be prepared by incorporating a small amount of polytetrafluoroethylene in the coating formulation. For example, a formulation of 100 parts of Ryton V-1, 33 parts of titanium dioxide, and 10 parts of polytetrafluoroethylene in aqueous slurry provides a coating that very readily releases food in a cooking operation. Contact angle, as measured with a Rame-Hart model A-100 goniometer, is 68° for Wesson oil and 110° for water on a coating surface of this type. A variety of colors may be obtained in release coatings by appropriate choice of pigment. The hardness of these coatings is somewhat greater than that of the conventional polytetrafluoroethylene cookware coatings (pencil hardness of 2H for the PPS coatings vs. H for polytetrafluoroethylene coating).

One very interesting application for polyphenylene sulfide is in coating cookware for nonstick use (10). Excellent, scratch-resistant, nonstick coatings are obtained when a formulation containing 10 to 20% polytetrafluoroethylene is used. Ryton PPS coatings are very insoluble and nontoxic, animal feeding studies indicate.

Table VII shows the effect of long-term aging of coatings at 500°F in air. The weight loss after 10 weeks exposure is less than 1% in the formulation containing a small amount of polytetrafluoroethylene.

Table VII.   Long-Term Thermal Stability in Air at 500° F.

	Weight Loss, %	
Exposure Time, Days	PPS/TiO₂ Coating (3/1)	PPS/TiO₂/PTFE Coating (3/1/0.3)
1	0.003	0.02
4	0.06	0.07
7	0.13	0.15
9	0.12	0.16
21	0.18	0.21
30	0.24	0.29
42	0.50	0.34
49	0.47 (Cracked)	0.31
70		0.95 (Cracked)

In other high-temperature tests, Ryton polyphenylene sulfide coatings on aluminum will pass an 80 inch-pound reverse impact test after exposure in air at 600°F for eight days, or at 700°F for two days (Table VIII).

Ryton polyphenylene sulfide coatings are finding excellent acceptance as corrosion-resistant, protective coatings for oil-field pipe, valves,

Table VII. Thermal Stability of Polyphenylene Sulfide Coatings [a]

	Evaluation [b]	
Exposure Conditions in Air	PPS/TiO₂ Coating (3/1)	PPS/TiO₂/PTFE Coating (3/1/0.3)
600°F—8 days	Pass	Pass
700°F—2 days	Pass	Pass

[a] One-mil coatings on aluminum
[b] As measured by 80 inch-pound reverse impact test

fittings, couplings, thermocouple wells, probes, and other equipment in both petroleum and chemical processing. Parts of this type have been operating satisfactorily for extended periods of time in media such as liquid ammonia, crude oil, refined hydrocarbons, motor oil, brine, dilute hydrochloric and sulfuric acids, dilute sodium hydroxide, butyl acetate, chlorobenzene, etc. In particular, polyphenylene sulfide is providing protection when both corrosive materials and high temperatures are involved. Thus, parts of carbon steel coated with various polyphenylene sulfide resins are replacing parts fabricated from expensive alloy metals. In many cases, corrosion-resistant coatings are applied by fluidized-bed techniques to obtain coating thicknesses of 10-25 mils readily. However, when the mass of the substrate metal is sufficient to hold the heat well, thick coatings may be obtained by spraying the hot (*ca.* 700°F) metal part with an aqueous coating slurry formulation, followed by a bake cycle. Both application methods are being used for corrosion protection.

Literature Cited

1. Duess, J. J. B., *Rec. Trav. Chim.*, (1908) **27**, 145.
2. Hilditch, T. P., *J. Chem. Soc.*, (1910) **47**, 2579.
3. Macallum, A. D., *J. Org. Chem.* (1948) **13**, 154; U.S. Patents **2,513,188** (June 27, 1950) and **2,538,941** (Jan. 23, 1951).
4. Lentz, R. W., and Carrington, W. K., *J. Polym. Sci.*, (1959) **41**, 333.
5. Lenz, R. W., and Handlovits, C. E., *J. Polym. Sci.*, (1960) **43**, 167.
6. Lenz, R. W., Handlovits, C. E., Smith, H. A., *J. Polym. Sci.*, (1962) **58**, 351.
7. Smith, H. A., *Encycl. Polymer Sci. Technol.*, (1969) **10**, 653.
8. Edmonds, Jr., J. T., Hill, Jr., H. W., U.S. Patent **3,354,129** (Nov. 21, 1967.)
9. Tabor, B. J., Magre, E. P., Boon, J., *Eur. Polym. J.*, (1971) **7**, 1127.
10. Ray, G. C., U.S. Patent **3,492,125** (Jan. 27, 1970).

RECEIVED April 14, 1972.

7

Polymerization of Cyclic Bis(arylene tetrasulfides)

NORMAN A. HIATT

Uniroyal Chemical, Division of
Uniroyal, Inc., Naugatuck, Conn. 06770

*Several cyclic bis(arylene tetrasulfides) have been synthe-
sized within the last few years. These ring compounds can
be polymerized thermally in much the same way as sulfur
itself. However, whereas polymeric sulfur depolymerizes
on standing at room temperature, the poly(arylene polysulfides) considered here are stable polymers under ordinary
conditions. The mechanism by which the cyclic bis-
(arylene tetrasulfides) polymerize has been established by
ESR as a free-radical process. The maximum number-
average molecular weights are from 15,000 to 16,000.
The polymers can function as tire-cord adhesives or as
metal-to-metal adhesives.*

There are very few examples in the literature of poly(arylene polysulfides). Perhaps the first such preparation was that of Friedel and
Crafts (*1*), in which benzene reacted with sulfur in the presence of
aluminum chloride. Within the last 15 years, several poly(arylene polysulfides) have been prepared by related reactions in which various aromatic compounds reacted with sulfur monochloride in the presence
of Friedel-Crafts catalysts (*2, 3, 4*). A variation of this reaction has
also been reported using a bifunctional sulfenyl chloride (*5*):

All of these Friedel-Crafts-type preparations may be classified as electrophilic substitutions.

The Macallum polymerization (*6, 7, 8*) for preparing poly(arylene polysulfides) has been shown by Lenz and co-workers (*9*) to involve a combination of free-radical and nucleophilic substitutions. This polymerization involves the reaction of a polyhaloaromatic compound with an alkali-metal sulfide or an alkaline-earth metal sulfide catalyzed by sulfur and carried out at a high temperature without solvent. Related to the Macallum reaction are two nucleophilic processes for the preparation of poly(phenylene sulfide). One of these, reported by Lenz and co-workers (*10, 11*), involves heating cuprous or sodium *p*-bromothiophenoxide at 250° to 305°C:

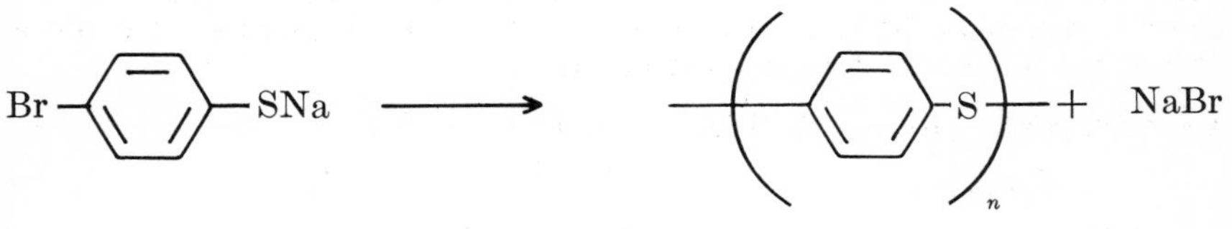

The other process involves the reaction of dichlorobenzene and sodium sulfide (*12*):

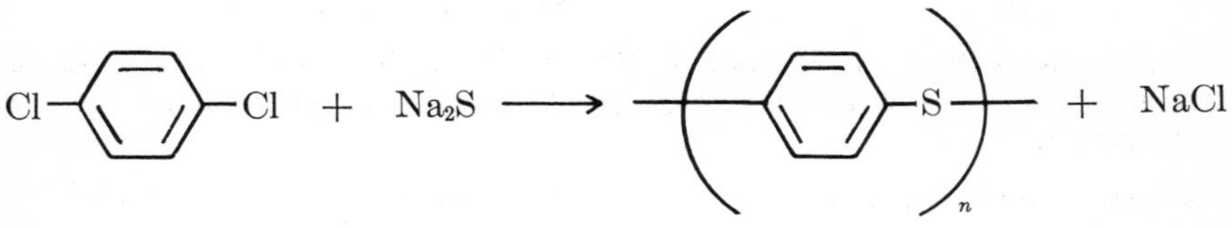

The work described here deals with a new approach to preparing poly(arylene polysulfide). This approach involves a free-radical mechanism—namely, the thermal polymerization of cyclic bis(arylene tetrasulfides):

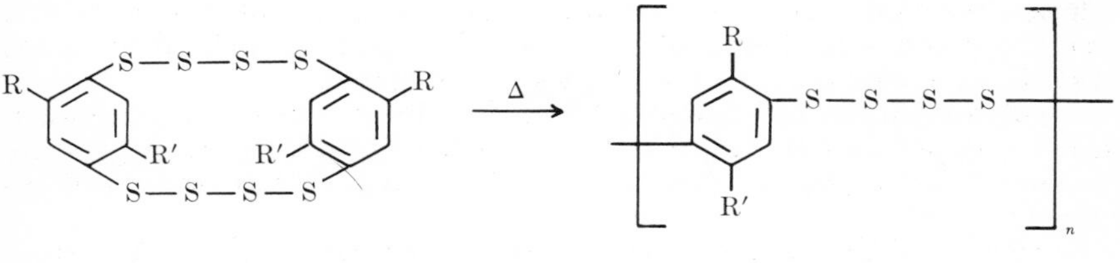

where   R = R′ = OEt   (Monomer   I)
        R = R′ = OMe   (Monomer II)
        R = OMe, R′ = Cl   (Monomer III)

These cyclic bis(arylene tetrasulfides) were originally synthesized by Z. S. Ariyan and R. L. Martin (*13, 14, 15*) by the reaction of the appropriate aromatic moiety with sulfur monochloride in the presence of a mineral acid-clay catalyst.

Experimental

Polymerization of Monomer I in Bulk (Optimum Conditions). Five grams of monomer I (*see* page 93, bottom, for structure) were placed in a test tube under a continuous stream of dry nitrogen. The tube was heated in an oil bath at 195°C for one hour. A viscous orange fluid was obtained; on cooling, the fluid solidified to a hard-orange glassy material. It was dissolved in 50 ml of chloroform and precipitated into 200 ml of methanol; the resulting yellow powdery polymer was dried *in vacuo* at 50°C for five hours. The yield of polymer was 4.8 g. The polymer softens at approximately 90°C.

Analysis: Calc. (for $C_{10}H_{12}S_4O_2$):   C 41.1   H 4.1   S 43.8

       Found: C 41.1   H 4.1   S 41.3

Polymerization of Monomer II in Bulk. The polymerization of monomer II was carried out in the same way as that of monomer I except that the temperature was maintained at 220° C.

Analysis: Calc. (for $C_8H_8S_4O_2$):   C 36.4   H 3.0   S 48.5

       Found: C 35.7   H 2.9   S 47.3

Polymerization of Monomer III in Bulk. The polymerization of monomer III was carried out under the same conditions as that of monomer I.

Analysis: Calc. (for $C_7H_5S_4ClO$):   C 31.3   H 1.9   S 47.6   Cl 13.2

       Found: C 30.0   H 1.9   S 45.8   Cl 12.8

Polymerization of Monomer I in Solution. Two grams of monomer I were dissolved in bromobenzene (2 g) and heated to 150°C with stirring under a continuous stream of dry nitrogen. The system was maintained at 150°C for three days, after which a viscous yellow-orange solution resulted. This solution was diluted with 20 ml of chloroform, and the polymer precipitated into 100 ml of methanol dried *in vacuo* at 50°C for five hours. A yield of 1.8 g was obtained.

Polymerization of Monomer I in a Hot Press. One gram of monomer I was placed between two Teflon-coated plates at contact pressure for one hour at 195°C. On cooling, an orange glassy film was obtained.

Attempted Copolymerization of Monomer I with Sulfur. A mixture of 2.5 g of sulfur and 2.5 g of monomer I was placed in a test tube under a continuous stream of dry nitrogen. The tube was heated in an oil bath at 185°C for two hours. A viscous orange fluid was obtained; on cooling, the fluid separated into two solid phases. One phase was a hard-orange glassy material, the other hard-yellow material. This heterogeneous mass was broken up and slurried in 50 ml of chloroform. The orange solid went into solution, and the yellow remained suspended in the chloroform. The system was then filtered and the yellow solid dried *in vacuo* at 50°C for five hours. A yield of 1.9 g was obtained. The melting point of this solid is 113-114°C, corresponding to that of sulfur. The orange material in chloroform solution was precipitated into 200 ml of methanol, and the resulting yellow powder dried *in vacuo* at 50°C for five hours. This material, 2.2 g, was identified as the polymer from monomer I.

Instruments used for the various physical and spectral measurements (with conditions, concentrations, and solvents noted where appropriate) were:

(a) Perkin Elmer ultraviolet-visible spectrophotometer, model 202 (solvent—CCl_4, conc.—0.02 g/l).

(b) Varian ESR spectrometer (250 gauss sweep in 5 min at 300 cm/hr); temp = 180°C; g = 2.023).

(c) Perkin Elmer differential scanning calorimeter, model 1B (scan rate = 20°C/min; temp range—270° to 370°K).

(d) Thomas melting-point apparatus.

(e) Cannon-Ubbelohde dilution viscometer (temp = 30°C; solvent —$CHCl_3$; conc—0.4 g/100 ml).

(f) Mechrolab vapor pressure osmometer (temp = 37°C; solvent— $CHCl_3$; conc range—40 to 102 g/l).

(g) Waters high-pressure liquid chromatograph, model ALC-202 (column—1 meter, 1/8 in. O.D., stainless-steel with 35-47μ Porasil T packing; mobile phase— 99.5%/0.5% of *n*-hexane/THF; flow rate = 1.0 ml/min; ambient temp.; UV detection).

Results and Discussion

General. Compounds such as monomers I, II and III have been considered as sulfur analogs of paracyclophane (*13*). However, their chemical behavior, at least with regard to polymerization, more closely resembles that of sulfur itself. For instance, paracyclophane is polymerized by vacuum distillation (*16*); sulfur (*17*) and monomers I, II, and III, by contrast, may be polymerized by heating in air or in an inert atmosphere.

Polymeric sulfur depolymerizes on standing after a short time (*18*). This is not the case with the poly(arylene polysulfides) which are stable indefinitely under ordinary conditions.

Molecular Weight. The maximum number-average molecular weights obtained thus far for the poly(arylene polysulfides) are from 15,000 to 16,000. Representative examples of the reduced viscosities and molecular weights, as determined by vapor pressure osmometry (VPO) in chloroform, are given in Table I.

Table I. Viscosity and Molecular Weights

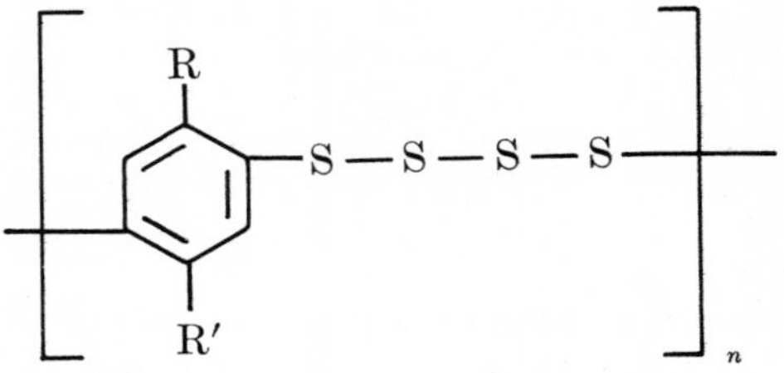

R	R'	Conditions of polymerization	Reduced viscosity (0.4 g/100 ml CHCl₃)	V.P.O. M.W.
OEt	OEt	1 hr @ 195° C (in bulk)	0.21	15,871
		3 days @ 150° C (in bromo-benzene)	0.08	7,129
		16 hr @ 150° C (in bromo-benzene)	0.06	4,653
OMe	OMe	1 hr @ 220° C (in bulk)	0.08	—
OMe	Cl	1 hr @ 195° C (in bulk)	0.04	—

The most likely explanation for the fact that higher-molecular-weight poly(arylene polysulfides) were not obtained probably can be traced to the purity of the monomer. An indication of this may be found with the high-pressure liquid chromatography conducted on monomer I. This method showed that although elemental sulfur was present in trace quantities (less than 0.1%), there was another (1 to 2%) unidentified impurity present despite repeated recrystallizations of the monomer. If this impurity were of the structure shown here, it could function as a chain-terminating agent under the conditions of the polymerization and thus lower the molecular weight.

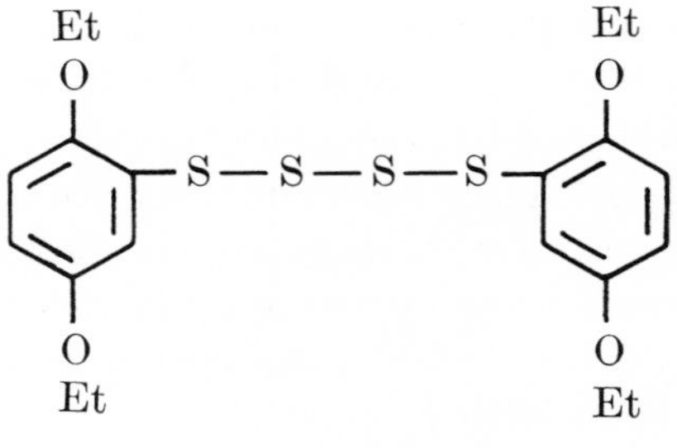

**Table II.   Sulfur Rank *vs.* Glass Transition Temperature
for Aromatic Polysulfides**

Structures	Average sulfur rank	T_g (° C)	T_m (° C)	M.W.
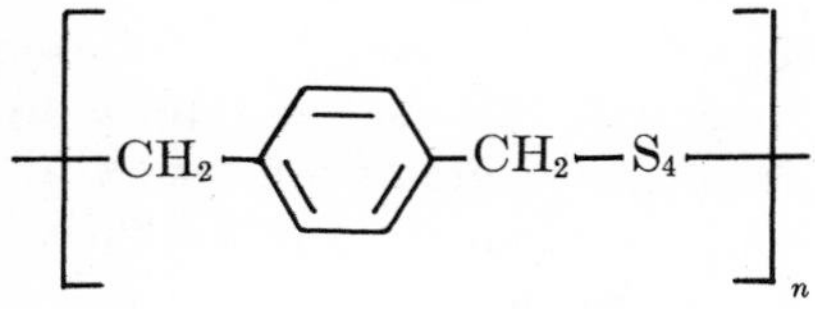	4	3 (*19*)	96 (*19*)	4,250 (*19*)
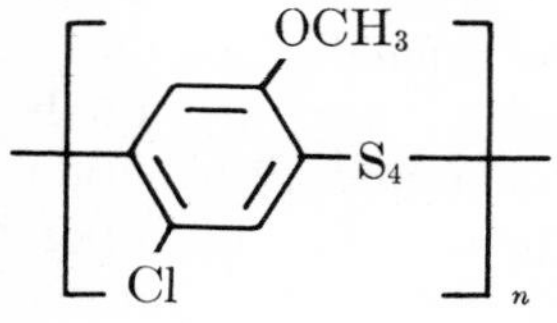	4	30	—	—
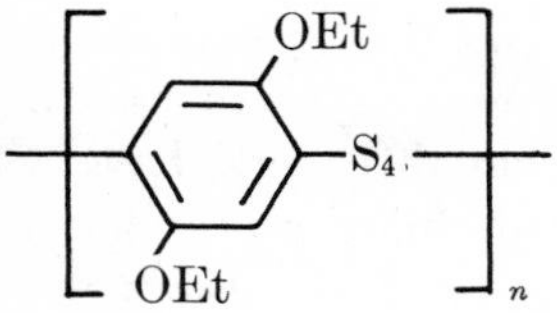	4	49	59	7,000–15,000
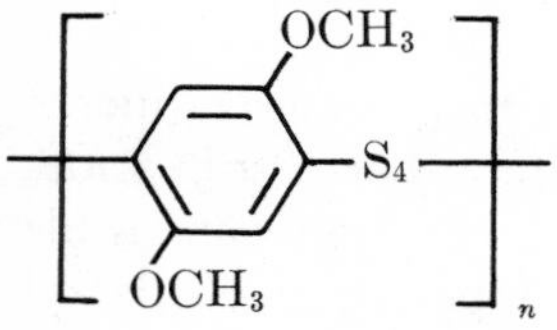	4	80	—	—
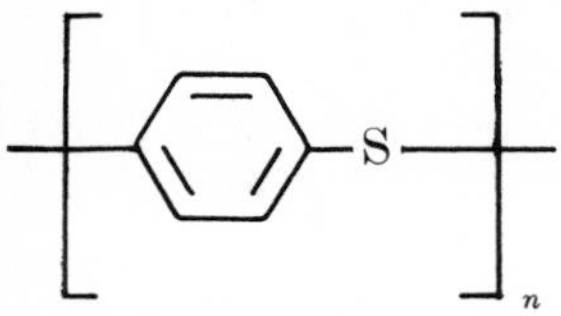	1	85 (*20*)	285 (*20*)	—

Another possible explanation might be that free sulfur splits out of the main chain to degrade the polymer. In regard to this, a small amount of elemental sulfur (less than 0.2%) sublimes out of the system during a bulk polymerization. High-pressure liquid chromatography conducted on the polymer formed from monomer I showed no residual monomer in the system and only trace quantities of elemental sulfur (less than 0.1%).

Attempts to copolymerize monomer I with sulfur resulted in formation of just the poly(arylene polysulfide) since all the sulfur separated out on cooling to room temperature. This is apparently because when eight or more sulfur atoms appear in sequence in a polymeric chain, the chain is unstable, and some of the sulfur eliminates to form the more stable eight-membered rings instead.

Sulfur Rank. In characterizing polysulfide polymers, it is customary to refer to the sulfur rank of the material. The sulfur rank is a numerical value designating the number of sulfur atoms appearing in sequence along the chain. In the most common commercial polysulfides, the Thiokol polymers, the sulfur rank is quite random along the polymeric chains as a result of the method used for their preparation; accordingly, an average sulfur rank is used. It is therefore difficult to correlate the properties of the polysulfide with the sulfur rank. Recently, Fitch and Helgeson (*19*) prepared polysulfides with a very narrow distribution of sulfur rank and were able to correlate this with the glass transition temperature, T_g. Generally, their conclusions were what one might expect: As sulfur rank increases, the T_g value decreases, and introduction of xylylene groups into the polysulfide chain increases the T_g value.

Aromatic polysulfides in which the sulfur atoms are attached directly to the aromatic ring, such as those studied in this work, have a substantially higher T_g value than the xylylene-containing polysulfides. Table II, which includes data on poly(phenylene sulfides), shows the relationships involving sulfur rank, ring substitution, and glass-transition temperature for aromatic polysulfides.

Polymerization by Ionic Catalysts. Attempts to polymerize monomer I with ionic catalysts such as trimethylphosphite or boron trifluoride-etherate at room temperature did not succeed. No reaction was visible in either case despite the fact that it has been reported that trialkyl phosphites react vigorously with sulfur at room temperature (*21*). However, even in that specific instance, no polymerization was observed. The use of an anionic initiator such as *n*-butyllithium did not produce polymer either. In this case, however, a reaction did occur, as evidenced by the discoloration of the monomer solution.

Polymerization by Irradiation. Using a Van de Graaff generator, solid samples of monomer I were irradiated at dose rates of from 4 to 256 watt-hours per pound. In addition, a sample dissolved in *o*-dichlorobenzene was irradiated at 256 watt-hours per pound. No polymerization occurred in any of the samples irradiated. However, the solid sample exposed at 256 watt-hours per pound changed color from a bright orange to a golden yellow, indicating that a reaction did take place.

Ultraviolet Spectra. During the course of their work, Z. S. Ariyan and R. L. Martin (*13*) determined that the UV absorption maximum of monomer I in chloroform occurs at λ_{max} of 371 mμ. This represents a bathochromic shift compared with other known tetrasulfides (290 to 330 mμ) (22). These data prompted them to speculate that there was internal steric strain in the ring caused by a nonbonding interaction between sulfur and oxygen:

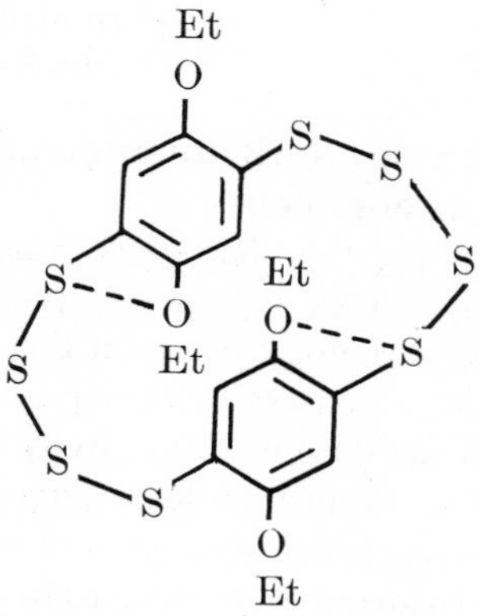

X-ray crystallographic studies on monomer I have tended to support S – – – O interactions (23).

In constructing a model of the polymer, we have observed that the sulfur and oxygen atoms no longer maintain the close proximity to each other as they do in the cyclic monomer. Therefore, the nonbonding interaction should diminish in importance. On this basis, we expected the UV absorption maximum of the polymer to be shifted to a lower wavelength than the monomer. This has been confirmed with the observation that the polymer obtained from monomer I showed a UV absorption maximum at 346.5 mμ.

Infrared and NMR Spectra. The infrared and NMR spectra of the poly(arylene polysulfides) are identical to those of the corresponding monomers. The major bands in the infrared for monomer I and the resulting polymer are:

(a) 3000 cm^{-1} (two peaks in this region, aromatic C–H and aliphatic C–H stretching);

(b) 1575 cm^{-1} (C=C in plane vibration);

(c) 1450 cm^{-1} and 1350 cm^{-1} (aliphatic C–H deformation); broad band in region of 1190 cm^{-1} (C–O stretching);

(d) 1025 cm^{-1} and 750 cm^{-1} (benzene-ring substitution).

The S–S and C–S absorption bands appear below 700 cm^{-1} and did not show up on the instrument used.

Absorption peaks in the NMR for monomer I and the resulting polymer are:

Type of proton	δ value (ppm)
CH₃ —	1.47 (triplet)
— CH₂ —	4.03 (multiplet)
ArH	7.23 (singlet)

Integration of areas under the NMR absorption peaks corresponds to the ratios of the different types of protons.

ESR Spectra. The question had been raised as to what type of polymerization mechanism was involved. We propose a free-radical process on the basis of the analogy between the cyclic bis(arylene tetrasulfides) and sulfur itself. To resolve this point, the polymerization of monomer I, in bulk, was carried out in the cavity of an ESR spectrometer. The ESR signal shown in Figure 1 clearly demonstrates a free-radical mechanism, and the g-value (spectroscopic splitting factor) of 2.023 for the free radicals formed during polymerization agrees closely with the g-value of 2.024 reported for polymerization of sulfur (24).

Figure 1.   ESR signal during polymerization of monomer I

Three steps are proposed as a plausible sequence in the polymerization mechanism:

(a) Initiation: S–S bond cleavage

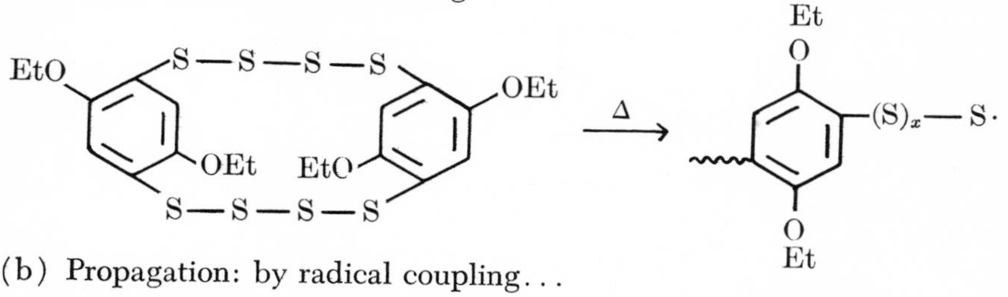

(b) Propagation: by radical coupling...

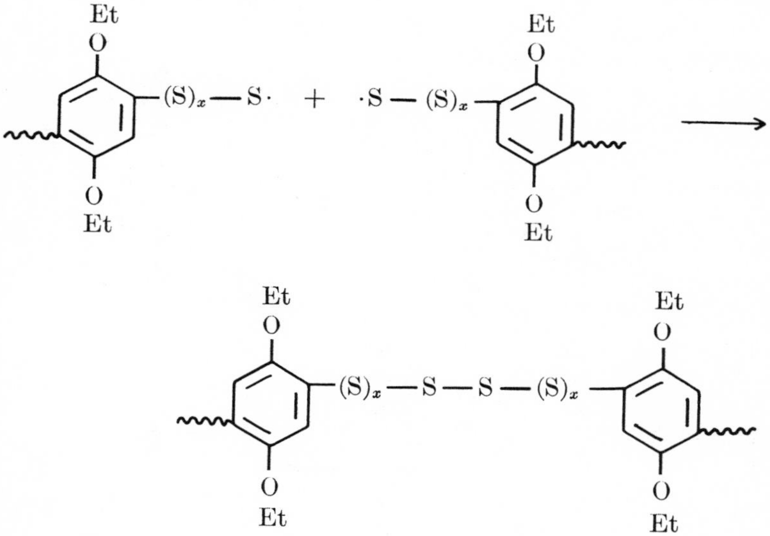

...and by attack of chain radical on monomer

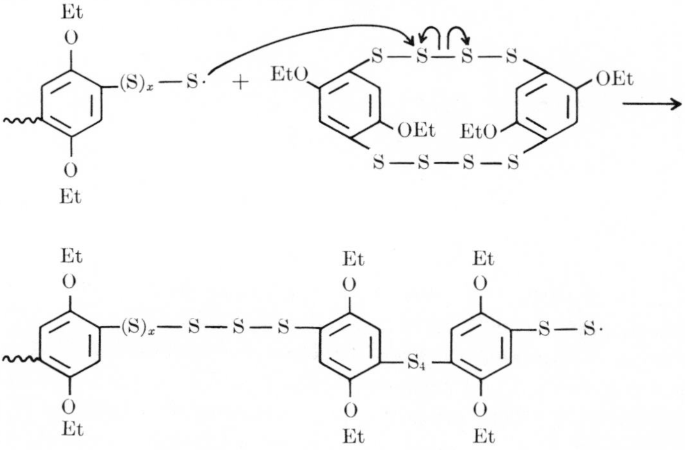

(c) Termination:

The presence of impurities such as...

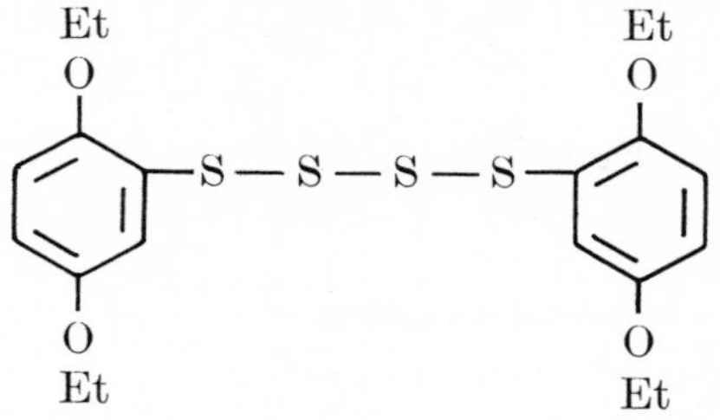

...could result in chain termination

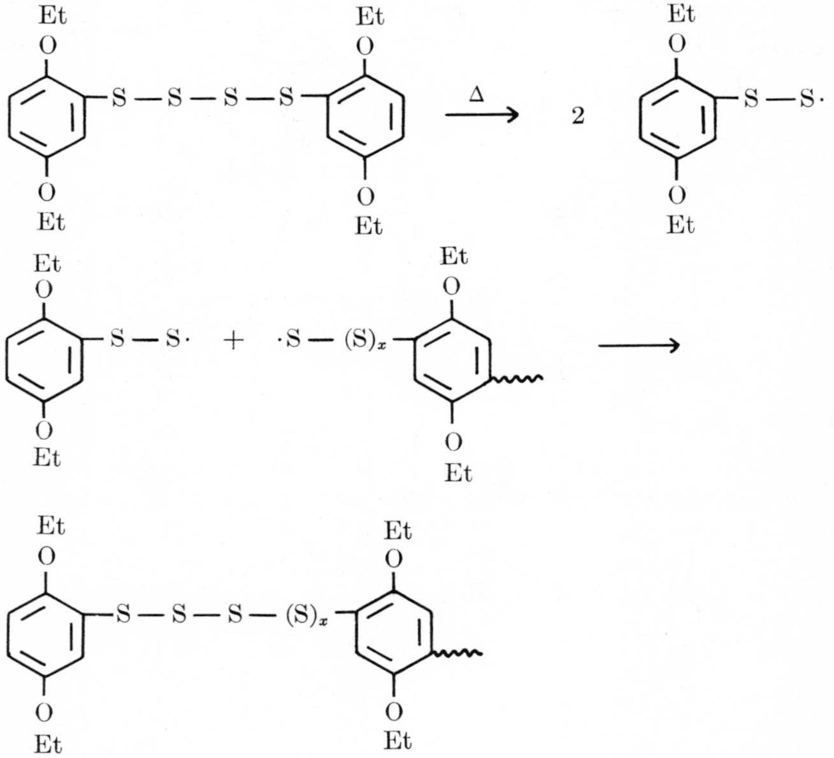

Similar initiation and propagation steps to those shown here have been proposed for the polymerization of sulfur (24).

Adhesive Properties of Poly(Arylene Polysulfides). The polymer obtained from monomer I has shown itself to be an adhesive for metal-to-metal bonding. This property was discovered when the monomer was polymerized in a hot press between two ferrotype plates. On cool-

Table III. Adhesive Tests for Polymer from Monomer I

(a) *Tire cord adhesion (ASTM D2229–63T)*
(*values are averages of three samples*)

Curing Time (min.)	Tensile Strength (psi)	Elongation (%)	Wire Adhesion % 250° F	Estimate of % Rubber Stock Left on Wire
15	2190	390	69	75
30	2000	250	69	75

(b) *Metal-to-metal adhesion (ASTM D2564–66T)*

Sample	Shear Strength (psi)
1	198
2	155
3	170

ing, the polymer adhered rather tenaciously to the plates.

The adhesive properties, coupled with the fact that monomer I was shown to be a vulcanizing agent for rubber (*15*), indicated that the polymer might make a suitable tire-cord adhesive. The data in Table III demonstrate the adhesive properties of the polymer, which may be data; and R. Kindle for the adhesion data. Irradiation work was done by D. I. Relyea at the Uniroyal Reseach Center in Wayne, N. J.

Acknowledgment

I wish to thank Z. S. Ariyan and W. Cummings for many helpful discussions; W. D. Spall for the high pressure liquid chromatography data; M. Tremeling for the E.S.R. spectra; A. Grant for the viscosity data; and R. Kindle for the adhesion data. Irradiation work was done by D. I. Relyea at the Uniroyal Research Center in Wayne, N. J.

Literature Cited

1. Friedel, C., Crafts, J. M., *Ann. Chim. Phys.* (1888) **6, 14,** 433.
2. Damanski, A. F., Binenfeld, Z., *Bull. Soc. Chim. Fr.* (1961) 679.
3. Degeorges, M. E., Bourgau, Y. (Progil S.A.) U.S.S.R. Patent **176, 553** (1965).
4. Fujisawa, T., Kakutani, M., *Polymer Letters* (1970) **8,** 19.
5. Fujisawa, T., Kakutani, M., *Polymer Letters* (1970) **8,** 511.
6. Macallum, A. D., *J. Org. Chem.* (1948) **13,** 154.
7. Macallum, A. D., U. S. Patent **2,513,188** (1950).
8. Macallum, A. D., U. S. Patent **2,538,941** (1951).
9. Lenz, R. W., Carrington, W. K., *J. Polym. Sci* (1959) **41,** 333.
10. Lenz, R. W., Handlovits, C. E., Smith, H. A., *J. Polym. Sci.* (1962) **58,** 351.

11. Lenz, R. W., Handlovits, C. E. Carrington, W. K., U. S. Patent **3,274,165** (1966).
12. Edmonds, J. T., Jr., Hill, H. W., Jr., U. S. Patent **3,354,129** (1967).
13. Ariyan, Z. S., Martin, R. L., *Chem. Commun.* (1969) 847.
14. Ariyan, Z. S., Martin, R. L., *Chem. Eng. News* (1969) **47** (42), 40.
15. Ariyan, Z. S., Martin, R. L., U. S. Patent **3,621,032** (1971).
16. Sorenson, W., Campbell, T., "Preparative Methods of Polymer Chemistry," 2nd edition, Interscience, New York, 1968.
17. Sorenson, W., Campbell, T., *Ibid.*
18. Tobolsky, A. V., MacKnight, W., "Polymeric Sulfur and Related Polymers," *Polymer Reviews* (1965) **13.**
19. Fitch, R., Helgeson, D., *J. Polym. Sci.*, (1969) *Pt. C* **22** 1101.
20. Short, J. N., Hill, H. W., Jr., *Chemtech* (1972) **2,** 481.
21. Bartlett, P. D., Meguerian, G., *J. Amer. Chem. Soc.* (1956) **78,** 3710.
22. Nakabayashi, T., Tsurugi, Jr., Takuzo, Y., *J. Org. Chem.* (1964) **29,** 1236.
23. Bernal, I., Ricci, J. S., *Chem. Commun.* (1969) 1453.
24. Fraenkel, G. K., Gardner, D. M., *J. Amer. Chem. Soc.* (1956) **78,** 3279.

RECEIVED June 1, 1972.

Poly(thiol esters)

HEINRICH G. BÜHRER and HANS-GEORG ELIAS

Midland Macromolecular Institute, 1910 West St. Andrews Dr., Midland, Mich. 48640, and Swiss Federal Institute of Technology, Universitätstrasse 6, CH-8006, Zürich, Switzerland.

The reported synthesis and properties of poly(thiol esters) of the A–B and A–A/B–B types are reviewed for these classes of compounds: 2 poly(α-mercapto acids), 7 poly(β-mercapto acids), 4 poly (ε-mercapto acids) and 44 A–A/B–B type poly(thiol esters). The melting points of poly(thiol esters) are generally lower than those of the corresponding polyamides and higher than those of the polyesters. They are readily hydrolyzed by alkali, which seems to be the main reason for the lack of commercial interest in these materials.

Poly(thiol esters) have been known for more than 25 years as stable, fiber-forming polymers; for a brief review *see* Goethals (*1*). Still, they have attracted little attention among scientists and have not been produced on a commercial scale. In this article, we summarize these often incomplete results to get a better understanding of this class of polymers.

In terms of their properties, poly(thiol esters) lie between polyesters and polyamides. In polyesters, oxygen atoms can be replaced formally by sulfur in three different ways to give poly(thiol esters) (I), poly(thion esters) (II), and poly(dithio esters) (III).

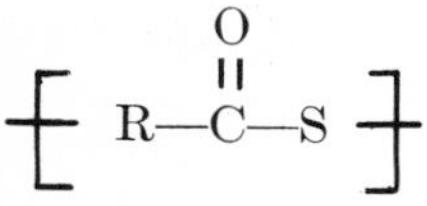

I 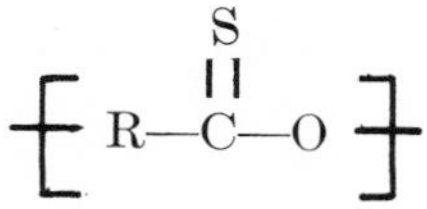 II 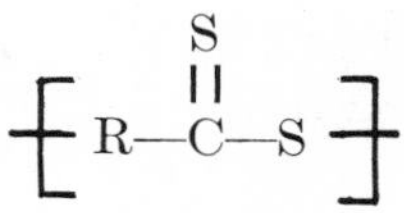 III

Although polymers of type II or III have not yet been made, it should be possible to synthesize them from the known poly(imido esters) (*2*) or poly(imidothiol esters) (*3*) by treatment with hydrogen sulfide.

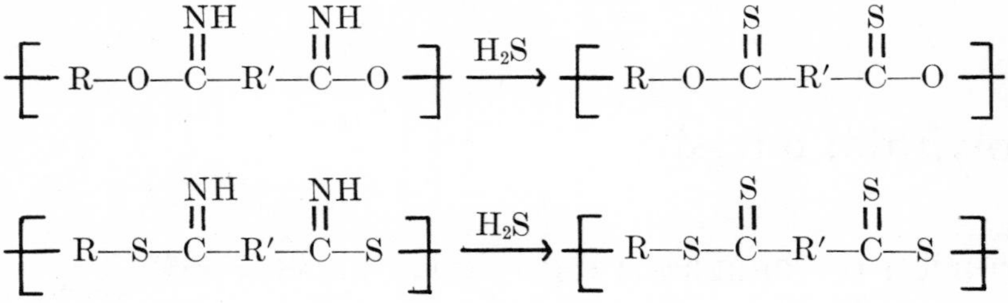

$$\left[\!\!\left[R\!-\!O\!-\!\overset{\overset{NH}{\|}}{C}\!-\!R'\!-\!\overset{\overset{NH}{\|}}{C}\!-\!O \right]\!\!\right] \overset{H_2S}{\longrightarrow} \left[\!\!\left[R\!-\!O\!-\!\overset{\overset{S}{\|}}{C}\!-\!R'\!-\!\overset{\overset{S}{\|}}{C}\!-\!O \right]\!\!\right]$$

$$\left[\!\!\left[R\!-\!S\!-\!\overset{\overset{NH}{\|}}{C}\!-\!R'\!-\!\overset{\overset{NH}{\|}}{C}\!-\!S \right]\!\!\right] \overset{H_2S}{\longrightarrow} \left[\!\!\left[R\!-\!S\!-\!\overset{\overset{S}{\|}}{C}\!-\!R'\!-\!\overset{\overset{S}{\|}}{C}\!-\!S \right]\!\!\right]$$

Poly(thiol esters) have been prepared by various methods, often analogous to preparation of polyesters, as is shown here.

As with polyamides, it is useful to distinguish between poly(thiol esters) of the A–B type and the A–A/B–B type. Typical examples are poly(ε-thiocaprolactone) (IV) and poly(hexamethylene dithiol terephthalate) (V).

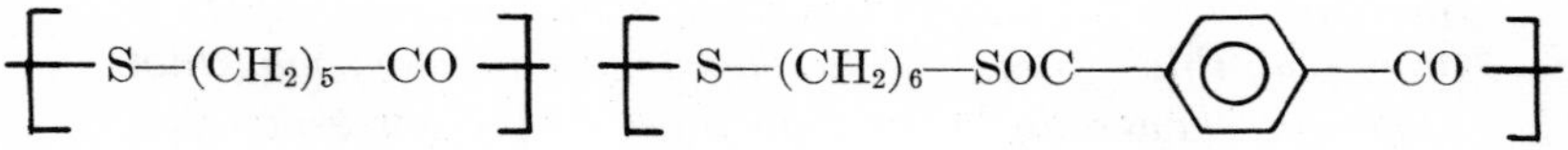

$$\left[\!\!\left[S\!-\!(CH_2)_5\!-\!CO \right]\!\!\right] \qquad \left[\!\!\left[S\!-\!(CH_2)_6\!-\!SOC\!-\!\!\bigcirc\!\!-\!CO \right]\!\!\right]$$

IV   V

Preparation of Poly(thiol esters) of the A–B Type

The most general preparation of these compounds is the polymerization of the corresponding thiolactones (VI).

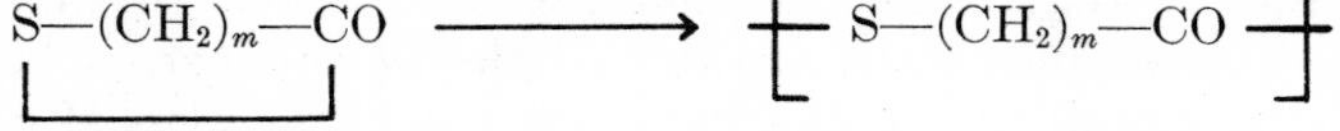

$$S\!-\!(CH_2)_m\!-\!CO \quad\longrightarrow\quad \left[\!\!\left[S\!-\!(CH_2)_m\!-\!CO \right]\!\!\right]$$

VI

Thiolactones and substituted thiolactones with four to seven ring atoms have been prepared. Except for γ-thiolactones, they have been polymerized. Other general procedures for making A–B type poly(thiol esters) are the internal addition reaction of ω-unsaturated thio acids (VII),

$$H_2C = CH\,(CH_2)_m\!-\!COSH \;\rightarrow\; \left[\!\!\left[S\!-\!(CH_2)_{m+2}\!-\!CO \right]\!\!\right]$$

VII

and the condensation polymerization of ω-mercapto acid chlorides bearing protective groups (such as benzyl, and trimethylsilyl) on the sulfur atom.

$$RS—(CH_2)_m—COCl \xrightarrow{-RCl} \left[S—(CH_2)_m—CO \right]$$

Poly(α-mercapto acids). The simplest compound of this type is poly(thioglycolide) (VIII)—poly(α-mercaptoacetic acid)—which has been prepared in different ways.

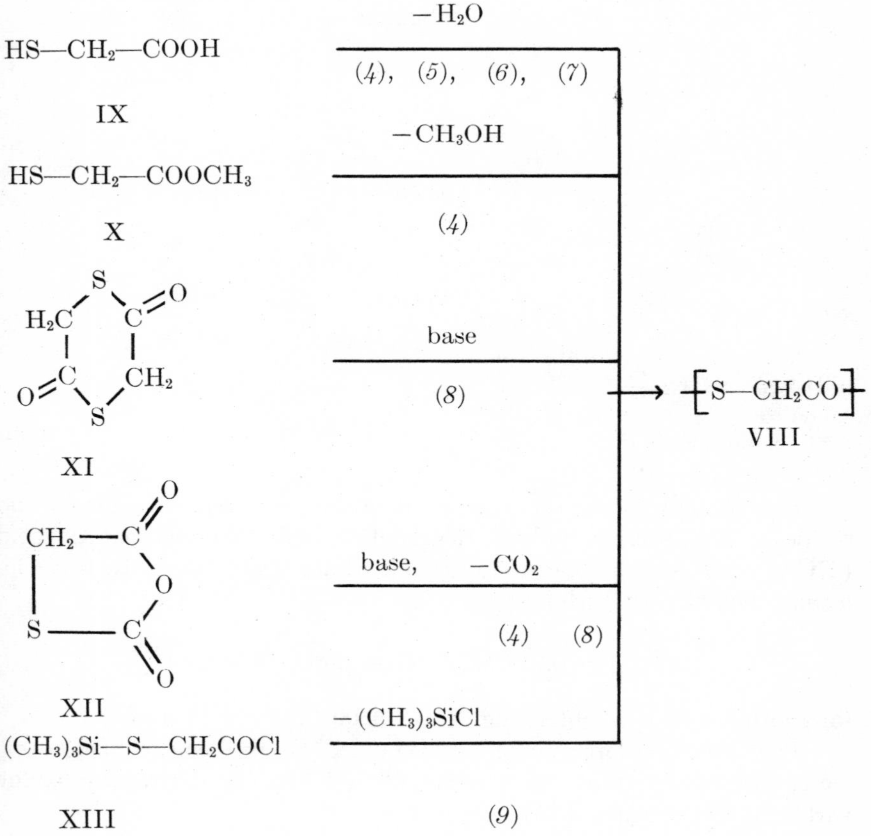

HS—CH$_2$—COOH

IX

HS—CH$_2$—COOCH$_3$

X

XI

XII

(CH$_3$)$_3$Si—S—CH$_2$COCl

XIII

$-H_2O$

(4), (5), (6), (7)

$-CH_3OH$

(4)

base

(8)

base, $-CO_2$

(4) (8)

$-(CH_3)_3SiCl$

(9)

$$\rightarrow \left[S—CH_2CO \right]$$
VIII

The condensation polymerization of IX and X yields oligomers only (*see* Table I). Condensation of IX gave dithioglycolide XI as a by-product (4). XI was also obtained by pyrolysis of poly(thioglycolide) (7). The polymerization of S-carboxy-α-mercaptoacetic acid anhydride XII using bases as initiators has been studied in detail (4). It is similar in many respects to the well-known polymerization of N-carboxy-α-amino acid anhydrides.

Table I.

Monomer	Solvent	Catalyst
HS—CH$_2$—COOH	(Bulk)	—
HS—CH$_2$—COOCH$_3$	(Bulk)	Ti(OBu)$_4$

| 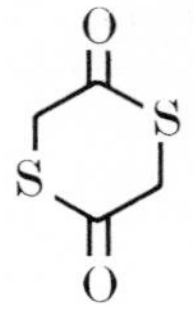 | Chloroform | Cyclohexylamine |

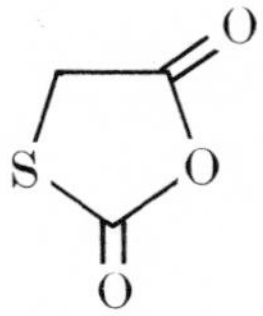	Dioxane	H$_2$O
"	Dioxane	Amines
"	Tetrahydrofuran	N(C$_2$H$_5$)$_3$
(CH$_3$)$_3$Si—S—CH$_2$—COCl	—	—

ᵃ In dioxane, 94.5°C

Poly(thioglycolide) is readily degraded by aqueous alkali and amines. It is soluble without degradation only in dichloroacetic acid (DCA) and hexafluoroacetone sesquihydrate (4). The viscosity-molecular weight relationship is

$$[\eta] = 0.018 \; \overline{M}_w^{0.72} \; ml/g \; (25°C, DCA)$$

for the molecular-weight range $\overline{M}_w$ of 5500 to 36000 Daltons (4).

Poly(α-mercaptopropionic acids) (XV), or poly(thiolactides), were synthesized from the corresponding S-Leuchs anhydrides having various optical purities (11).

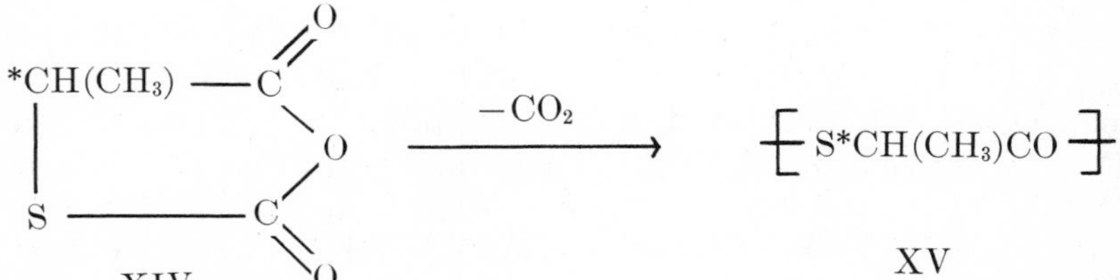

Poly(thioglycolides)

Temp., °C	Yield, %	$[\eta]$, ml/g	$\bar{x}_n$	mp, °C	Reference
130	100	—	9	Liquid	(7)
140	—	—	3	"	(4)
25	94	—	—	165–169	(8)
100	97.4	15 ^{a}	45	140–145	(10)
25	100	—	—	147–157	(8)
−50	93.2	32.6 ^{b}	440 ^{c}	157	(4)
—	—	50 ^{b}	—	—	(9)

^{b} In dichloroacetic acid, 25°C
^{c} $\bar{x}_w$ (from light scattering)

The polymerization in dioxane solution with amines as initiators was very slow and incomplete (*see* Table II). However, a rapid solid-state polymerization was observed during the attempted sublimation or recrystallization of the monomer XIV.

Poly(β-mercapto acids). Poly(β-mercaptopropionic acid) (XVI) has been prepared this way (12) (*see also* Table III):

$$HSCH_2CH_2COOH + Cl\overset{O}{\overset{||}{C}}OEt + NEt_3$$

$$\downarrow$$

$$[HSCH_2CH_2CO{-}O{-}\overset{O}{\overset{||}{C}}OEt] + NEt_3H^+Cl^-$$

$$\downarrow \Delta$$

$$\bigg[SCH_2CH_2CO\bigg] \quad + \quad EtOH \quad + \quad CO_2$$

XVI

An alpha-substituted poly(β-mercaptopropionic acid) has been

Table II.   Poly(thiolactides)

Monomer	Solvent	Catalyst	Temp., °C	Yield, %	$[\eta]_{sp}/c$, ml/g	mp, °C	Reference
$D_{29}L_{71}$ 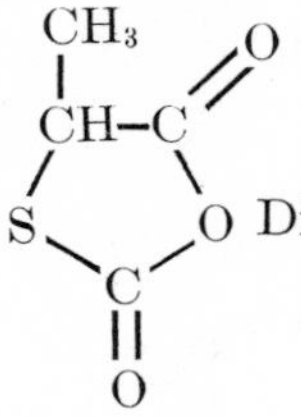 (see structure)	Dioxane	n-Hexyl-amine	25	16	37.7^a	133	(11)
$D_{50}L_{50}$	"	"	"	"	—	17.4^b	$(^c)$	(11)
$D_{93}L_7$	"	"	H_2O	"	—	—	152	(11)

^{a} In chloroform, 25°C; $c = 0.02$ g/ml
^{b} In chloroform, 25°C; $c = 0.03$ g/ml
^{c} Amorphous; glass-transition point at 12°C

made by treating S-benzyl-N-phthalyl-L-cysteinyl chloride (XVII) with one equivalent of $AlCl_3$ (13).

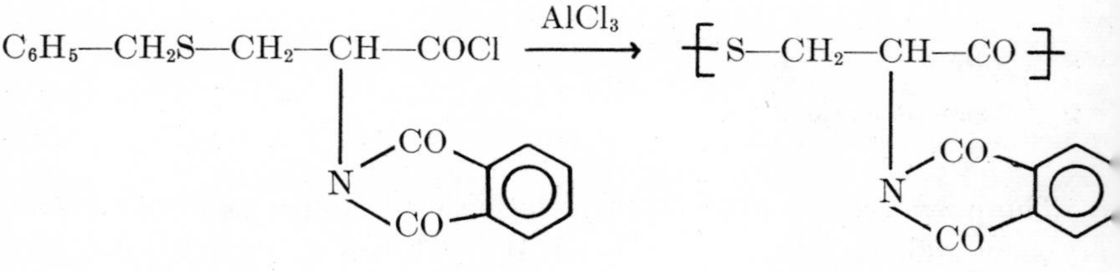

XVII

When two equivalents of $AlCl_3$ were used, the corresponding β-thio propiolactone was obtained. A similar product was made by ring opening polymerization of (S)-(−)-α-p-toluenesulfonamido-β-thiopropiolactone (XVIII) (14).

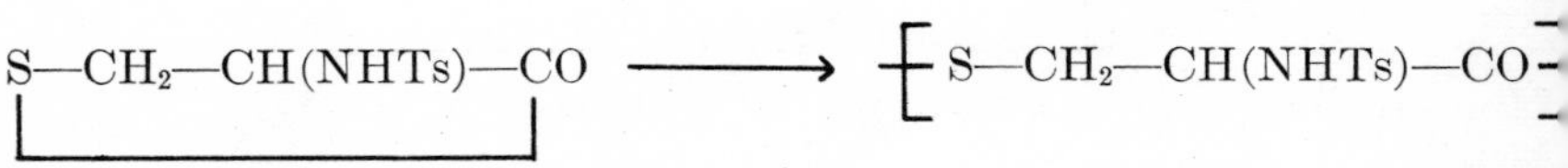

XVIII

Water, benzylmercaptan, and amines were used as initiators. The polymers had number-average degrees of polymerization up to 50.

U.S. Patent 3,367,921 (16) describes the polymerization of some β-thiolactones of type XIX, where R and R′ are methyl or ethyl.

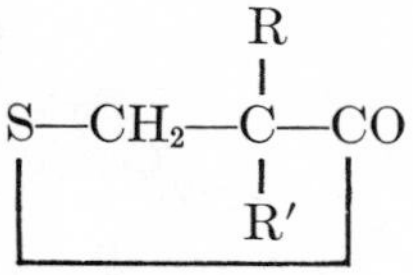

$$
\begin{array}{c}
\text{R} \\
| \\
\boxed{\text{S—CH}_2\text{—C—CO}} \\
| \\
\text{R}'
\end{array}
\qquad \text{XIX}
$$

The polymerization with strong bases yielded high-molecular-weight poly(thiol esters), which could be extruded to films and fibers. Some poly(thiol ester-coesters) are also described in the patent.

β,β-Dimethyl-β-thiopropiolactone (XX) was polymerized with a trace of water (*15*).

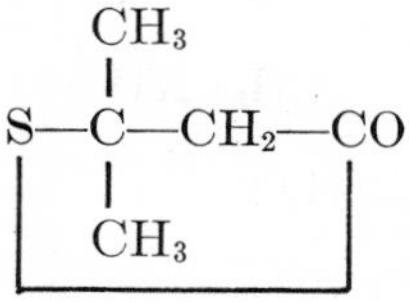

$$
\begin{array}{c}
\text{CH}_3 \\
| \\
\boxed{\text{S—C—CH}_2\text{—CO}} \\
| \\
\text{CH}_3
\end{array}
\qquad \text{XX}
$$

No polymer properties were reported.

Poly(γ-mercapto acids). γ-Thiobutyrolactone did not polymerize in bulk at 155°C with potassium *tert*-butoxide as initiator (*17*). This is probably because of an unfavorable monomer-polymer equilibrium, resulting in a low ceiling temperature. It might be possible to overcome these difficulties by using lower polymerization temperatures or high pressure techniques, or both. A similar behavior has been reported for γ-butyrolactone (*18*), which could be polymerized at elevated temperatures at 20,000 atm. It is likely that substituted γ-thiolactones are even more difficult to polymerize.

Poly(δ-mercapto acids). δ-Thiovalerolactone was polymerized in 21% yield to a linear poly(thiol ester)—$[\eta]^{25}_{\text{CHCl}_3} = 14.7$ ml/gram, m.p. 117°-118°C (*17*). The molecular weight of the polymer seems to be controlled by the monomer–polymer equilibrium. The same product (XXII) was obtained by distilling thiol-4-pentenoic acid (XXI) (*19*), giving the thiolactone (XXIII) as a by-product.

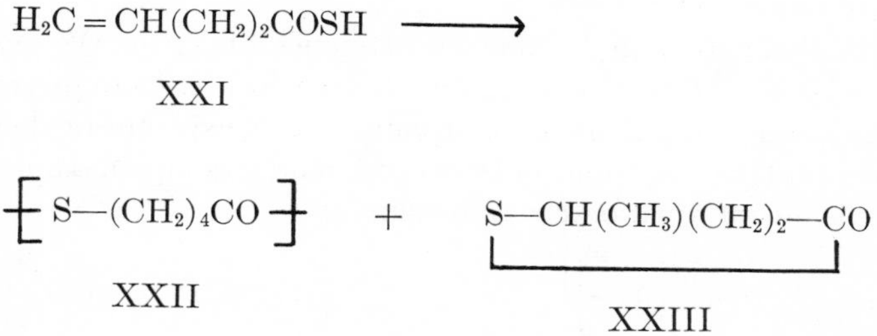

$$
\text{H}_2\text{C} = \text{CH(CH}_2)_2\text{COSH} \longrightarrow
$$

$$
\text{XXI}
$$

$$
\left\{ \text{S—(CH}_2)_4\text{CO} \right\} \quad + \quad \boxed{\text{S—CH(CH}_3)\text{(CH}_2)_2\text{—CO}}
$$

$$
\text{XXII} \qquad\qquad\qquad\qquad \text{XXIII}
$$

The structure of XXII was proved by alkaline hydrolysis, which gave

Table III

Monomer	Solvent	Catalyst
HS—CH$_2$CH$_2$—COOH	Ether	N(C$_2$H$_5$)$_3$/ClCOOC$_2$H$_5$

Monomer	Solvent	Catalyst
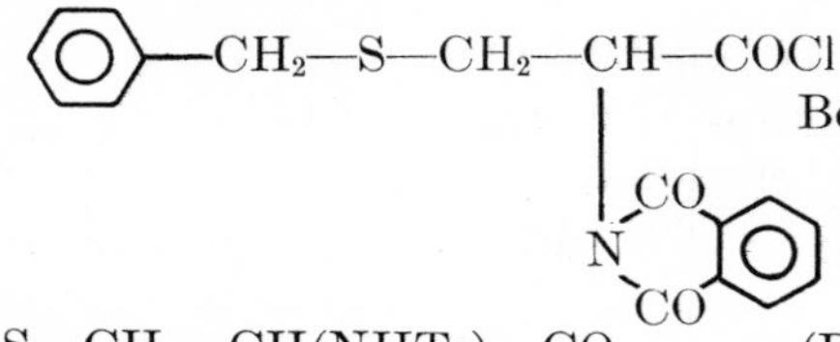	Benzene	Stoichiometric amounts of AlCl$_3$
S—CH$_2$—CH(NHTs)—CO	(Bulk)	Thermally
"	DMF	(CH$_3$)$_2$NH
S—C(CH$_3$)$_2$—CH$_2$—CO	(Bulk)	H$_2$O
S—CH$_2$—C(CH$_3$)$_2$—CO	Benzene	Base
S—CH$_2$—C(C$_2$H$_5$)$_2$—CO	"	"
S—CH$_2$—C(CH$_3$)(C$_2$H$_5$)—CO	"	"

[a] In DMF, 30°C

δ-mercaptovaleric acid. The addition thus did not follow Markownikoff's rule. The melting point of XXII ($\bar{x}_n = 9$) was 115°–117°C.

Poly(ε-mercapto acids). In a series of papers, Overberger and Weise (17, 20, 21) reported the polymerization of ε-thiocaprolactone (XXIV) and of some substituted analogs (see Table VI). XXIV could be polymerized in bulk or tetrahydrofuran solution, using bases such as butyllithium, potassium tert-butoxide, or sodium as initiators. AlCl$_3$ as initiator gave a crosslinked product. Linear poly(ε-thiocaprolactone) is crystalline and soluble in chlorinated hydrocarbons.

The same product was obtained by intermolecular olefin addition of thiol-5-hexenoic acid (19), but melting point and degree of polymerization were much lower.

The methyl-substituted ε-thiocaprolactones of Table IV were polymerized in bulk. (R)-(−)-poly(γ-methyl-ε-thiocaprolactone) was crystalline, the other two polymers amorphous. ORD spectra of these polymers were similar to low-molecular-weight model compounds, indicating that no helical conformation exists in solution.

$$\left[\!\!\left[S—(CH_2)_6—CO \right]\!\!\right] \qquad \left[\!\!\left[S—(CH_2)_{10}—CO \right]\!\!\right]$$

XXV   XXVI

Poly(β-mercapto acids)

Temp., °C	Yield, %	[η], ml/g	$\bar{x}_n$	mp, °C	Reference
25	—	—	—	134–145	(12)
65	31	—	6	256–260	(13)
102	—	20.6 ^{a}	33	237–240	(14)
27	—	28.5 ^{a}	49	245–250	(14)
—	—	—	—	—	(15)
100	85	1.34 ^{b}	—	—	(16)
"	68	0.85 ^{b}	—	—	(16)
"	78	1.02 ^{b}	—	—	(16)

^{b} = {η} in toluene, 30°C [g/dl]

Poly(ω-mercapto acids). Poly(ω-mercaptoheptanoic acid) (XXV) — ($\bar{x}_n$ = 11, m.p. = 69°C) (19)—and poly(ω-mercaptoundecylenic acid) (XXVI)—$[\eta]^{25}_{CHCl_3}$ = 26 ml/g (22)—were prepared from the ω-unsaturated thioacids.

The low molecular weights obtained were attributed to the fact that the very reactive monomers could not be purified by distillation, the impurities acting as chain terminators.

Preparation of Poly(thiol esters) of the A–A/B–B Type

The methods by which these poly(thiol esters) have been synthesized are summarized in Table V. Three other types of reactions exist that could also yield poly(thiol esters):

HSOC—R—COSH + HS—R'—SH ⇌ ┤OC—R—COS—R'—S├ + H_2S

O=C=CH—R—CH=C=O + HS—R'—SH →
┤OCCH$_2$—R—CH$_2$COS—R'—S├

NaSOC—R—COSNa + Br—R'—Br → ┤SOC—R—COSR'├ + 2 NaBr

These reactions have not been used so far as we know although they give thiolesters with monofunctional reactants (23, 24, 25).

Table VI gives the poly(thiol esters) made by the reactions mentioned in Table V. In several cases, information given about polymer

Table IV.

Monomer	Solvent	Catalyst
$H_2C=CH-(CH_2)_3-COSH$	(Bulk)	Spontaneous
$S-(CH_2)_5-CO$ ⎿_____⏌	"	*tert*-BuOK
"	Tetrahydro-furan	*n*-BuLi
$R-(-)$ $S-(CH_2)_3-CH(CH_3)-CH_2-CO$ ⎿_____⏌	(Bulk)	*tert*-BuOK
$(\pm)$ $S-(CH_2)_2-CH(CH_3)-(CH_2)_2-CO$ ⎿_____⏌	"	*n*-BuLi
$R-(+)$ "	"	"

a In chloroform, 25°C

Table V.   Types of Reactions Leading to A–A/B–B Type Poly(thiol esters)

Reactant 1	Reactant 2	Type	Abbreviation in Table VI
$HOOC-R-COOH$	$HS-R'-SH$	condensation polymerization	A/M
$R''OOC-R-COOR''$	$HS-R'-SH$	condensation polymerization	E/M
$ClOC-R-COCl$	$HS-R'-SH$	condensation polymerization	C/M
$HSOC-R-COSH$	$CH_2=CH(CH_2)_2-CH=CH_2$	Adduct formation	T
$HSOC-R-COSH$	$CH\equiv C-(CH_2)_3-CH_3$	Adduct formation	T

properties is very incomplete. The reaction types shown in Table V are dealt with here.

Poly(thiol esters) from Dibasic Acids and Dimercaptans (**Reaction A/M**). By this procedure (26), only low molecular weights were obtained because of side reactions. The reaction has therefore not found wider application, compared with the synthesis of polyesters.

From Dibasic Acids Esters and Dimercaptans (Reaction E/M). This transesterification (31) offers some advantages over other methods.

Poly(ε-mercapto acids)

Temp., °C	Yield, %	[η], ml/g	x̄ₙ	mp, °C	Tg[b], °C	Reference
—	92	—	16	80–83	—	(19)
155	78	76.3 [a]	—	104–6	—	(20)
25	86	55.8 [a]	—	102–4	19	(17)
110–190	76	54.2 [a]	—	Amorphous	−50, +20	(21)
25	55	34.1 [a]	—	"	—	(21)
"	28	42.1 [a]	—	62	—	(21)

[b] Glass-transition temperature by DTA

The diphenyl esters used are easily purified, and the reaction can be catalyzed by lithium, lithium hydride, dibutyltin oxide, and other compounds. By performing the process in two stages (first at 200°C at atmospheric pressure followed by 320°C under high vacuum), it is possible to obtain the high molecular weights required for fiber-forming properties. Diphenyl esters of dibasic acids with the phenol substituted by electron-acceptor groups show increased reactivity; alphatic esters gave no polymers.

From Dibasic Acid Chlorides and Dimercaptans (Reaction C/M). This is the most widely used reaction to produce poly(thiol esters) of the A–A/B–B type (27, 28, 29, 32, 33, 34). The polymerization process can be carried out three different ways:

(a) Condensation in bulk (BC).

(b) Condensation in solution (SC).

(c) Interfacial condensation (IC).

A number of poly(thiol esters) have been synthesized by method BC (29). The reaction temperature was increased from room temperature to above 200°C. The products were often colored because of side reactions (replacement of hydroxyl with chlorine, formation of ketene structures, and ester pyrolysis are known side reactions in polyester formation).

In method SC, polymers often precipitate during the reaction. Molecular weights of products produced by this reaction are lower than those obtained by the other two methods (*see* Table VI). In some cases, pyridine is added to the solution (28).

Method IC permits using lower reaction temperatures. Thus,

Table VI.

Structural Unit	Reaction Type	Leaving Group	Solvent
$-S(CH_2)_2SOC-CH=CH-CO-$	A/M	H_2O	Benzene
$-S(CH_2)_2SOC(CH_2)_4CO-$	C/M	HCl	(interfacial)
"	C/M	"	Benzene
$-S(CH_2)_2SOC(CH_2)_8CO-$	C/M	"	(interfacial)
"	"	"	—
$-S(CH_2)_4SOC(CH_2)_4CO-$	"	"	(interfacial)
"	"	"	Benzene
"	"	"	(Bulk)
$-S(CH_2)_4SOC(CH_2)_8CO-$	"	"	"
$-S(CH_2)_6SOC(CH_2)_4CO-$	"	"	"
"	C/M	HCl	Benzene
"	TA/U	—	Benzene
"	"	—	(Emulsion)
"	"	—	"
$-S(CH_2)_6SOC(CH_2)_5CO-$	TA/U	—	Benzene
"	"	—	(Emulsion)
"	"	—	"
$-S(CH_2)_6SOC(CH_2)_6CO-$	"	—	Benzene
"	"	—	Tetrahydrofuran
$-S(CH_2)_6SOC(CH_2)_6CO-$	"	—	(Emulsion)
"	"	—	"
$-S(CH_2)_6SOC(CH_2)_7CO-$	"	—	Benzene
"	"	—	(Emulsion)
"	"	—	"
$-S(CH_2)_6SOC(CH_2)_8CO-$	C/M	HCl	—
"	TA/U	—	Benzene
"	"	—	(Emulsion)
"	"	—	"

Poly(thiol esters)

Catalyst	Temp., °C	Yield, %	$\{\eta\}$, g/dl	$\eta_m{}^b$	mp, °C	Reference
HCl	Reflux	—	—	—	—	(26)
NaOH	—	80	0.5[a]	—	125	(27)
Pyridine	Reflux	62	0.13[c]	—	111–125	(28)
Base	25	69	0.4[a]	—	—	(27)
—	110	—	—	1.6	Liquid	(29)
NaOH	25	—	0.5[a]	—	100	(27)
Pyridine	Reflux	50	0.13[c]	—	95–100	(28)
—	218	—	—	19000	125–128	(29)
—	218	—	—	1870	95–98	(29)
—	218	—	—	720	113–115	(29)
Pyridine	Reflux	51	0.15[c]	—	97–104	(28)
UV	—	72	0.16[c]	—	99–102	(28)
Persulfate/bisulfite	30	100	0.32[c]	—	—	(30)
H₂O₂	30	87	0.50[c]	—	100	(30)
UV	—	61	0.19[c]	—	61–64	(28)
Persulfate/bisulfite	30	84	0.35[c]	—	—	(30)
H₂O₂	30	74	0.31[c]	—	70	(30)
UV	—	69	0.12[c]	—	82–85	(28)
UV	—	24	0.36[c]	—	—	(30)
Persulfate/bisulfite	30	93	0.33[c]	—	—	(30)
H₂O₂	30	91	0.92[c]	—	110	(30)
UV	—	75	0.21[c]	—	68–72	(28)
Persulfate/bisulfite	30	99	0.52[c]	—	—	(30)
H₂O₂	30	79	1.0[c]	—	105	(30)
—	218	—	—	216	107–109	(29)
UV	—	65	0.17[c]	—	68–80	(28)
Persulfate/bisulfite	30	96	0.52[c]	—	—	(30)
H₂O₂	30	89	1.3[c]	—	140	(30)

Table VI.

Structural Unit	Reaction Type	Leaving Group	Solvent
—S(CH$_2$)$_{10}$SOC(CH$_2$)$_4$CO—	C/M	HCl	Benzene
—S(CH$_2$)$_{10}$SOC(CH$_2$)$_8$CO—	"	"	—
—S(CH$_2$)$_2$S(CH$_2$)$_2$SOC(CH$_2$)$_4$CO—	"	"	Benzene
Polymer from HSOC(CH$_2$)$_8$COSH + 1-hexyne	TA/U	—	(Emulsion)
 —S—(H)—SOC(CH$_2$)$_2$CO—	E/M	Phenol	(Bulk)
 —SCH$_2$—(H)—CH$_2$SOC—(H)—CO—	"	o-Cresol	"
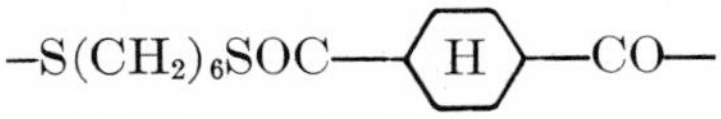 —S(CH$_2$)$_6$SOC—(H)—CO—	"	Phenol	"
—S(CH$_2$)$_2$SOC—(O)—CO—	C/M	HCl	(interfacial)
"	"	"	Benzene
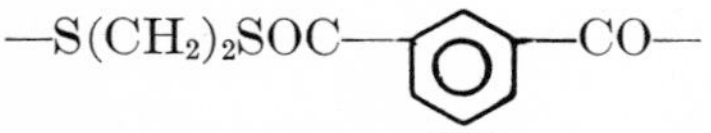 —S(CH$_2$)$_2$SOC—(O)—CO—	"	"	(Interfacial)
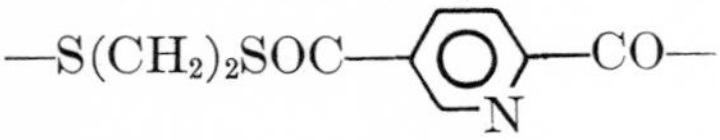 —S(CH$_2$)$_2$SOC—(O)$_N$—CO—	"	"	"
—S(CH$_2$)$_3$SOC—(O)—CO—	"	"	(Bulk)
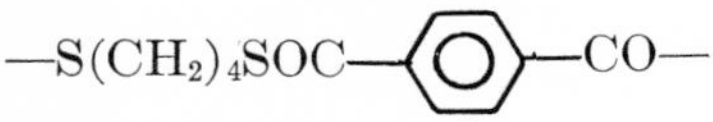 —S(CH$_2$)$_4$SOC—(O)—CO—	"	"	"
"	"	"	Benzene
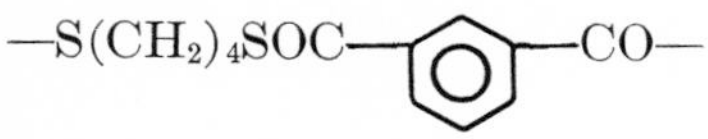 —S(CH$_2$)$_4$SOC—(O)—CO—	E/M	Phenol	(Bulk)
"	C/M	HCl	"
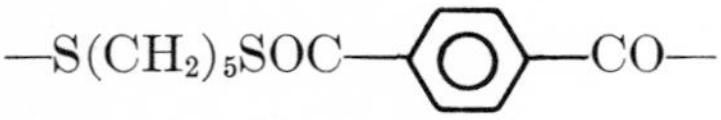 —S(CH$_2$)$_5$SOC—(O)—CO—	"	"	"
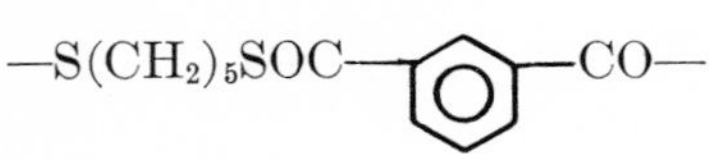 —S(CH$_2$)$_5$SOC—(O)—CO—	"	"	(Interfacial)
"	"	"	(Bulk)
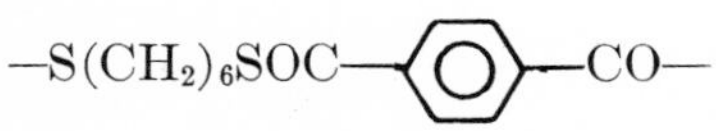 —S(CH$_2$)$_6$SOC—(O)—CO—	E/M	Phenol	"
"	C/M	HCl	(Interfacial)
"	TA/U	—	Benzene

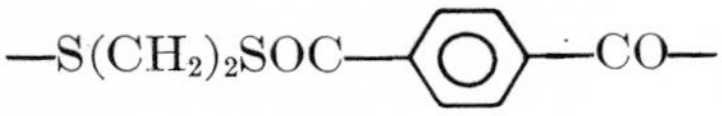

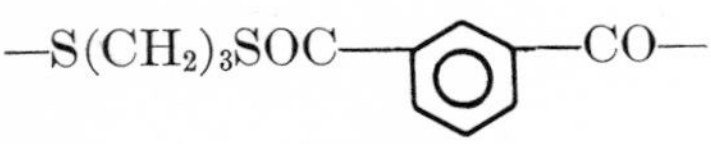

Continued

Catalyst	Temp., °C	Yield, %	$\{\eta\}$, g/dl	$\eta_m{}^b$	mp, °C	Reference
Pyridine	Reflux	57	0.10[c]	—	70–75	(28)
—	218	—	—	2300	100–103	(29)
Pyridine	Reflux	47	0.12[c]	—	67–70	(28)
H_2O_2	30	60	0.33[c]	—	—	(22)
LiH	250	—	0.68[d]	—	291–302	(31)
"	250	—	—	—	302–310	(31)
"	"	—	—	—	188–215	(31)
NaOH	15	100	—	—	340[e]	(27)
Pyridine	Reflux	68	—	—	335–340	(28)
NaOH	50	77	0.35[d]	—	185[e]	(27)
"	25	—	—	—	280[d]	(32)
—	218	—	—	10,000	([d])	(29)
—	340	—	—	—	310	(29)
Pyridine	Reflux	20	—	—	230–260	(28)
LiH	250	—	0.53[d]	—	168–174	(31)
—	218	—	—	198	162–165	(29)
—	255	—	—	3500	231–232	(29)
NaOH	25	78	0.57[a]	—	175[e]	(27)
—	218	—	—	386	104–106	(29)
CaH_2	250	—	0.83[d]	—	280–300	(31)
NaOH	55	100	0.41[d]	—	260–270	(27)
UV	—	92	0.07[c]	—	160–200	(28)

Table VI.

Structural Unit	Reaction Type	Leaving Group	Solvent
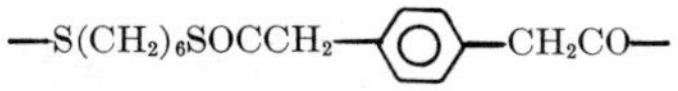 —S(CH₂)₆SOCCH₂—◯—CH₂CO—	"	—	(Emulsion)
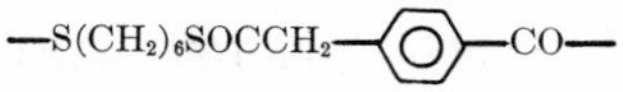 —S(CH₂)₆SOCCH₂—◯—CO—	"	—	"
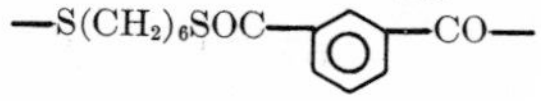 —S(CH₂)₆SOC—◯—CO—	C/M	HCl	(Bulk)
"	TA/U	—	Benzene
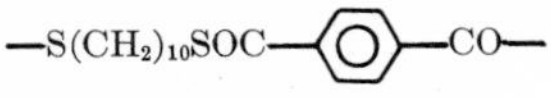 —S(CH₂)₁₀SOC—◯—CO—	C/M	HCl	(Bulk)
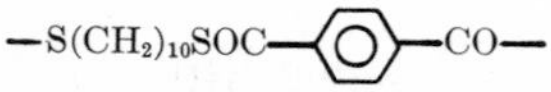 —S(CH₂)₁₀SOC—◯—CO—	C/M	HCl	Benzene
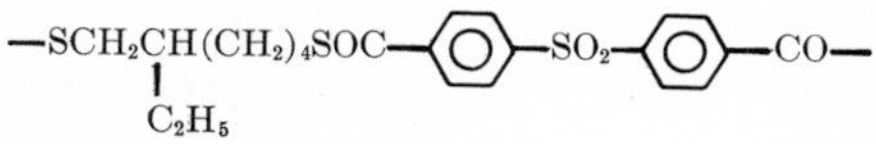 —SCH₂CH(CH₂)₄SOC—◯—SO₂—◯—CO— (C₂H₅)	E/M	o-Cl-phenol	(Bulk)
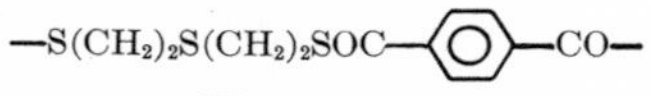 —S(CH₂)₂S(CH₂)₂SOC—◯—CO—	C/M	HCl	Benzene
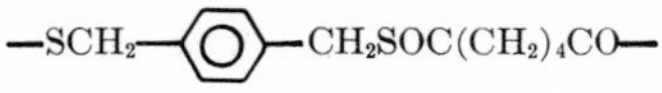 —SCH₂—◯—CH₂SOC(CH₂)₄CO—	"	"	(Bulk)
"	"	"	Xylene
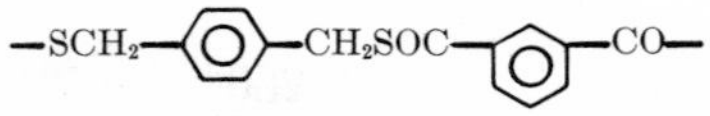 —SCH₂—◯—CH₂SOC—◯—CO—	"	"	(Bulk)
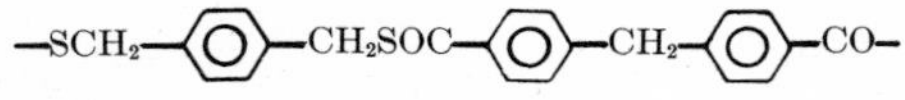 —SCH₂—◯—CH₂SOC—◯—CH₂—◯—CO—	E/M	p-Cresol	"
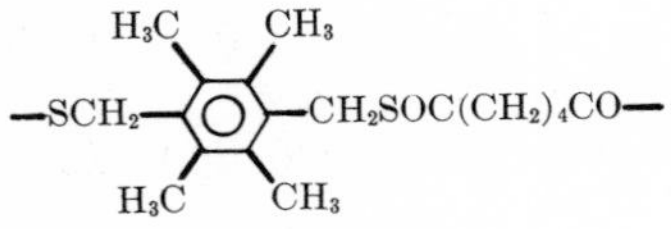 —SCH₂—(H₃C,CH₃,H₃C,CH₃)—CH₂SOC(CH₂)₄CO—	C/M	HCl	Xylene
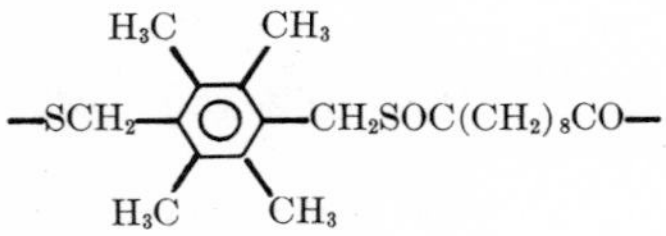 —SCH₂—(H₃C,CH₃,H₃C,CH₃)—CH₂SOC(CH₂)₈CO—	"	"	"
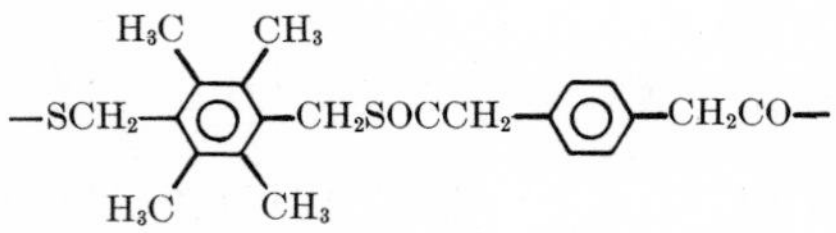 —SCH₂—(H₃C,CH₃,H₃C,CH₃)—CH₂SOCCH₂—◯—CH₂CO—	E/M	Phenol	(Bulk)
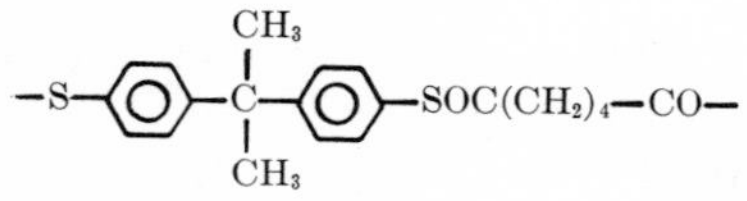 —S—◯—C(CH₃)(CH₃)—◯—SOC(CH₂)₄—CO—	"	2,6-diCl-phenol	(Bulk)
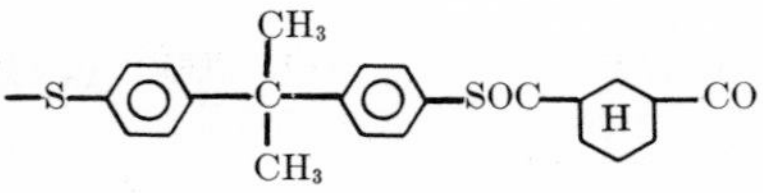 —S—◯—C(CH₃)(CH₃)—◯—SOC—◯(H)—CO	E/M	Phenol	(Bulk)

Continued

Catalyst	Temp., °C	Yield, %	$\{\eta\}$, g/dl	$\eta_m{}^b$	mp, °C	Reference
H_2O_2	30	96	0.71^c	—	105	(30)
"	30	38	0.16^c	—	$(^f)$	(30)
—	218	—	—	660	110–113	(29)
UV	—	22	0.07^c	—	75–82	(28)
—	218	—	—	3290	200–201	(29)
Pyridine	Reflux	71	—	—	136–144	(28)
LiH	230	—	0.56^d	—	200–208	(31)
Pyridine	Reflux	73	—	—	140–146	(28)
—	218	—	—	1360	168–170	(29)
—	Reflux	—	—	—	190	(33)
—	237	—	—	785	200–210	(29)
LiH	240	—	0.89^d	—	268–274	(31)
—	Reflux	100	—	—	255	(33)
—	"	—	—	—	180	(33)
LiH	265	—	0.61^d	—	257–262	(31)
"	250	—	—	—	185–198	(31)
LiH	260	—	0.51^d	—	205–220	(21)

Table VI.

Structural Unit	Reaction Type	Leaving Group	Solvent
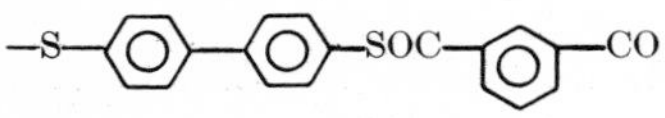 —S—⟨O⟩—⟨O⟩—SOC—⟨O⟩—CO	C/M	HCl	(Interfacial)
"	"	"	dimethyl-acetamide
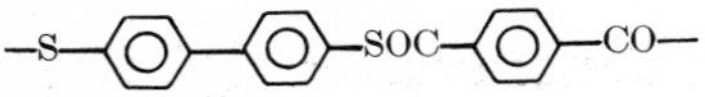 —S—⟨O⟩—⟨O⟩—SOC—⟨O⟩—CO—	"	"	"
"	"	"	dichloro-ethane

[a] In *m*-cresol, 25°C
[b] Melt viscosity at reaction temperature (poise)
[c] In chloroform, 25°C
[d] In phenol–tetrachloroethane (60/40)
[e] "Sticking" temperature

isocinchomeron acid dichloride (XXVII) reacts with dimercaptoethane at room temperature without decomposition of the unstable pyridine compound (32).

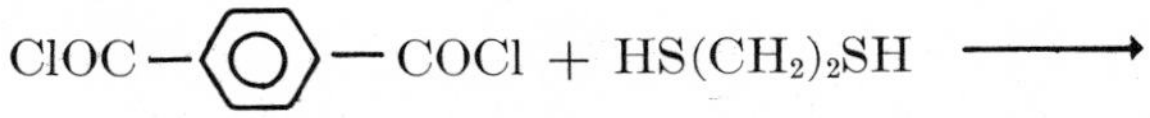

$$ClOC-\langle O \rangle-COCl + HS(CH_2)_2SH \longrightarrow$$

XXVII

$$\{S(CH_2)_2SOC-\langle O \rangle-CO\} \quad + 2\ HCl$$

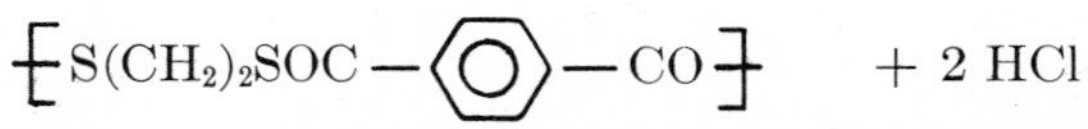

Some all-aromatic poly(thiol esters) have also been prepared by this method (34). There is also a continuous process for interfacial condensation polymerization by which poly(thiol esters) with high inherent viscosities have been produced (27).

From Dithio Dibasic Acids and Unsaturated Compounds (Reaction TA/U). Poly(thiol esters) from dibasic thio acids and biallyl were first prepared in benzene or chloroform solution in the presence of UV light (28). Molecular weights of the products obtained were low. The reaction product from biallyl and dithioladipic acid had the same infrared pattern as that from adipyl chloride and hexamethylenedithiol, proving the anti-Markownikoff character of the adduct formation. Use of an emulsion polymerization technique gave higher molecular weights (30). Dithiolterephthalic acid and dithiolisophthalic acid gave no polymers with biallyl. Instead, the acids were oxidized to the polymeric acid disulfides (XXVIII) (22).

The molecular weights of the poly(thiol esters) increased with decreasing water solubility of the dithiolcarboxylic acid. Initiation with hydrogen peroxide gave higher molecular weights than did initiation

Continued

Catalyst	Temp., °C	Yield, %	$\{\eta\}$, g/dl	$\eta_m{}^b$	mp, °C	Refer-ence
NaOH	25	72	104[g]	—	330–340	(34)
—	25	75	11[h]	—	330–340	(34)
NEt₃	25	97	24[i]	—	>460	(34)
"	25	60	40[k]	—	—	(34)

[f] Gummy
[g] η_{spec}/c, 25°C, tetrachloroethane-phenol, $c = 0.0014$ g/ml
[h] Same as [g] $c = 0.0044$ g/ml
[i] Same as [g] $c = 0.0006$ g/ml
[k] Same as [g] $c = 0.0005$ g/ml

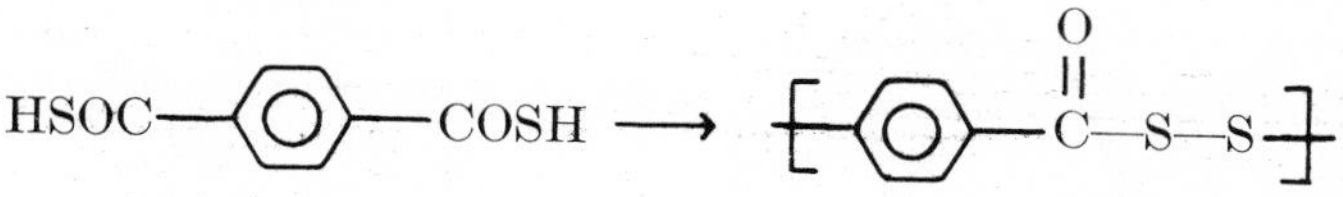

$$HSOC—\bigcirc—COSH \longrightarrow \left[\bigcirc—\overset{\overset{\displaystyle O}{\|}}{C}—S—S \right]$$

XXVIII

with ammonium persulfate–sodium metabisulfite.

The addition product of dithiolsebacic acid and 1-hexyne was also prepared (22), but its structure is unknown.

Properties and Applications

High-molecular-weight poly(thiol esters) show fiber-forming properties. This section covers the available data concerning melting points, crystallinity, solubility, and stability of this class of esters, and compares them with polyesters and polyamides. Information about spectra is also reviewed.

Melting Points. The melting points of the various polymers are given in Figures 1 through 4 (more data are in Tables I-IV and VI). In these formulas, X may be O, S, or NH.

$$\left[X(CH_2)_mCO \right]_n \qquad \left[X(CH_2)_6XOC(CH_2)_mCO \right]_n$$

$$\left[X(CH_2)_mXOC(CH_2)_4CO \right]_n$$

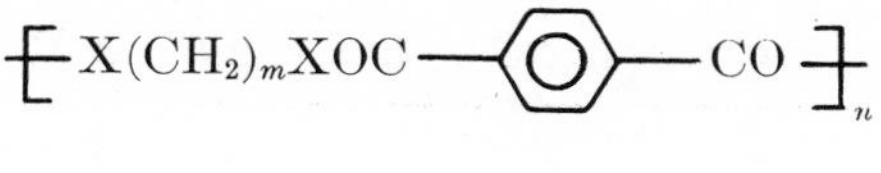

$$\left[X(CH_2)_mXOC—\bigcirc—CO \right]_n$$

Melting points of polyesters and polyamides were taken from the Brandrup, J., Immergut, E. H., "Polymer Handbook," Wiley, New York, 1967. The data do not represent true thermodynamic melting points but have been determined for polymers of unknown degree of crystallinity by methods such as differential thermal analysis, polarization microscopy, capillary tube, and others. Five conclusions can be drawn from the data.

(1) The melting points of poly(thiol esters) lie between those polyesters and polyamides (Figures 1-4). They melt 30° to 70°C higher than do the corresponding polyesters, the shape of the curves being very similar. This behavior is most likely because of a decreased flexibility of the $-\overset{O}{\overset{\|}{C}}-S-$ bond, compared with the $-\overset{O}{\overset{\|}{C}}-O-$ bond. The only exceptions are poly(thioglycolide), which melts 60°C below poly(glycolide) (Figure 1), and poly(decamethylene dithioladipate) (Figure 3). The lower melting point of poly(thioglycolide) compared with poly(glycolide) is probably caused by the helical conformation of the latter.

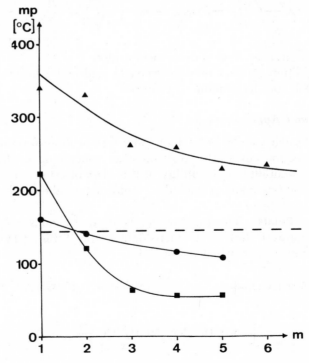

Figure 1.   Melting points of polymers with the monomer unit $-[X(CH_2)_mCO]-$; ▲ $X = NH$; ■ $X = O$; ● $X = S$;—m.p. of polyethylene

The low melting point of poly(decamethylene dithioladipate) may be caused by the low molecular weight ($[\eta]^{25}_{CHCl_3} = 10$ g/ml).

   (2) The melting points of aliphatic poly(thiol esters) (Figures 2 and 3) lie below that of polyethylene (143°C) and tend to increase with

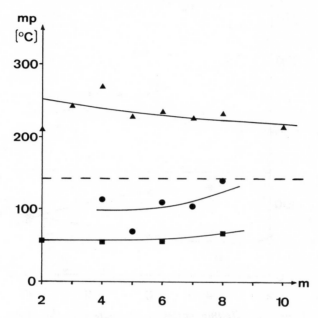

Figure 2.   Melting points of polymers with the monomer unit $+X(CH_2)_6XOC(CH_2)_mCO+$; ▲ $X = NH$; ■ $X = O$; ● $X = S$;—*m.p. of polyethylene*

increasing number of methylene groups per monomer unit.   This is not true for some poly(thiolactones); *see* Figure 1.

   (3) The introduction of alicyclic parts into the poly(thiol ester) chain increases conformational rigidity and, consequently, the melting point.   This can be seen from the melting points of poly(hexamethylene

$+S(CH_2)_6SOC(CH_2)_4CO+_n$

XXIX

113°–115°C

$+S(CH_2)_6SOC-\langle H \rangle-CO+_n$

XXX

188–215°C

$+SCH_2-\langle H \rangle-CH_2SOC-\langle H \rangle-CO+_n$

XXXI

302–310°C

dithioladipate)(XXIX), poly(hexamethylene dithiol-*trans*-cyclohexane-1,
4-dicarboxylate) (XXX), and poly(1,4-dimethylene-*trans*-cyclohexane
dithiol-*trans*-cyclohexane-1,4-dicarboxylate)(XXXI).

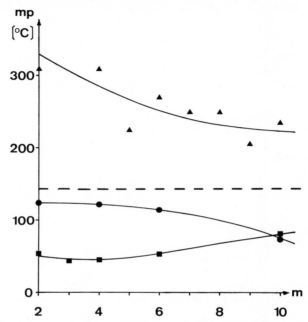

*Figure 3. Melting points of polymers with the mono-
mer unit* ⊦X(CH₂)ₘXOC(CH₂)₄CO⊣, ▲ X = NH; ■ X
= O; ● X = S;—m.p. of polyethylene

In XXXI, when the *trans*-dimercaptan is replaced by a 70:30 mixture of
trans and cis isomers, the melting point is lowered to 240°–255°C.

(4) Poly(thiol esters) with a terephthalic moiety in the chain
(Figure 4) have melting points higher than polyethylene. These melting points decrease with increasing number of methylene groups in the
monomer unit. Poly(thiol esters) with terephthalic units always melt at
higher temperatures than do the isophthalic isomers (Table VI).

(5) All-aromatic poly(thiol esters), such as poly(4,4'-biphenylene
dithiolterephthalate)(XXXII) have very high melting points (*34*).

$$\left[\text{S}-\langle\bigcirc\rangle-\langle\bigcirc\rangle-\text{SOC}-\langle\bigcirc\rangle-\text{CO}\right]_n$$

XXXII

>460°C

From these results, it is evident that a number of aromatic or alicyclic
poly(thiol esters) have melting points suitable for making fibers.

Very few glass-transition temperatures have been published (*see*
Table IV).

Crystallinity. Most poly(thiol esters) are quite crystalline. Crystallinity can be lowered by using nonlinear monomers—for example, $HSCH_2(C_2H_5)CH–(CH_2)_4SH$ (*31*)—or by copolymerization (*29*). The x-ray pattern of a uniaxially oriented poly(ε-thiocaprolactone) (*17*) indicates that this polymer adopts a planar structure with extended chains.

Solubility. The sparse data available show no marked difference between polyesters and poly(thiol esters) in their behavior toward organic solvents. Good solvents for many poly(thiol esters) are chlorinated hydrocarbons, phenol, and organic acids such as dichloroacetic acid. With the exception of poly(thioglycolide) (*4*) and poly(thiolactide) (*11*), the problem of degradation by organic solvent action was not studied systematically. Solubility is increased in poly(thiol ester coesters) (*34*).

Stability. Only few studies of the stability of poly(thiol esters)

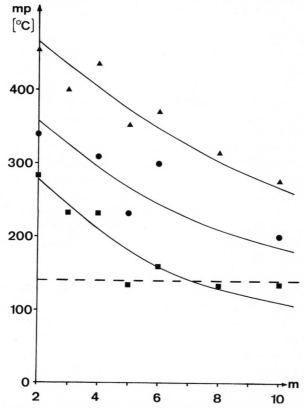

Figure 4. Melting points of polymers with the monomer unit $+X(CH_2)_mXOC$ —⬡— $CO+$; ▲ $X = NH$; ■ $X = O$; ● $X = S$; —*m.p. of polyethylene*

against chemical agents have been made. It appears that the esters are very stable against hydrolysis in water or acids but readily react with aqueous alkali (28). Aminolysis with primary and secondary amines is very fast and yields the low-molecular-weight amides. Poly(thioglycolide) is also degraded by dimethylformamide and tertiary amines (4).

Applying results from low-molecular-weight thiolesters and esters to the corresponding polymers should give slightly slower acid hydrolysis, equally fast alkaline hydrolysis, and a much faster aminolysis of poly-(thiol esters) compared with polyesters (24, 35, 36).

No data are available concerning the thermal stability of poly(thiol esters).

Infrared Spectra. The infrared spectra of poly(thiol esters) (4, 11,

17, 28) show the characteristic $-\overset{\overset{\displaystyle O}{\|}}{C}-$ stretch at 1675–1700 cm^{-1}.

Bands at 900–1000 cm^{-1} are attributed to the $-\overset{\overset{\displaystyle O}{\|}}{C}-S-$ stretch vibration,

and those at 1100–1200 cm^{-1} to the $-C-\overset{\overset{\displaystyle O}{\|}}{C}-$ stretch vibration (37).

UV and ORD Spectra. The UV spectra of thiol esters have a peak at about 230 nm (log $\varepsilon \approx 3.6$) (38) ascribed to a transition of the $\pi \to \pi^*$ type. Similar absorption maxima were found in poly(thiol esters) (17 and 21). In addition, in the UV spectrum of poly(thio-lactide) (XV), a shoulder at 279 nm (log $\varepsilon = 2.5$ in CHCl$_3$) (11) may be the result of a $n \to \pi^*$ type absorption.

ORD and CD measurements of optically active thiolacetates (39), poly(thiolactides) (XV) (11), and poly(ε-thiolactones) (21) show two Cotton effects, one centered around 230–240 nm and the other at about 280 nm. The ORD spectra of the polymers are nearly identical with those of low-molecular-weight model compounds, thus indicating the absence of a helical conformation of the polymers in solution.

Applications. Poly(thiol esters) can be good starting materials for producing fibers when the esters have an inherent viscosity $\{\eta\} > 0.3$ g/dl (27) or >0.75 g/dl (16). They can be cold-drawn to oriented fibers involving a three- to fivefold or even higher increase in length.

Poly(α,α-dimethyl-β-thiopropiolactone) has been melt-spun at 185°C to give a fiber which, after drawing, had a tenacity of 1.4 g/den. (polyethylene terephthalate, 4–7 g/den.) and an initial modulus of 12 g/den. (30–130 g/den.) (16). Tensile recovery at 10% elongation was 80%. No information is available about poly(thiol esters) with higher melting points, such as poly(hexamethylene dithiolterephthalate).

Films can be cast from poly(thiol esters) with inherent viscosities $\{\eta\} > 0.5$ g/dl (16), but no data about their properties are available.

Moderately crystalline poly(thiol esters) may prove valuable in producing molding plastics.

The fast reaction between poly(thiol esters) and amines has been used to increase the sulfur content of casein- and lysine-containing polypeptides (*40*).

Conclusions

The main differences between poly(thiol esters) on the one hand and polyamides and polyesters on the other may be summarized this way:

(a) Poly(thiol ester) synthesis is more complicated and expensive since reactions using diacids directly as monomers are not possible; it is necessary to use either the acid chlorides or the phenol esters.

(b) Dimercaptans are more expensive than glycols or diamines.

(c) The main advantage of poly(thiol esters) over polyesters is the former's higher melting point.

However, polymers prepared from thiolactones melt at a temperature too low for fiber use.

Considering these points, it is understandable why poly(thiol esters) have never attracted commercial interest.   Furthermore, most poly(thiol esters) seem to be unstable against alkalis.   The odors of the by-products of hydrolysis and thermal degradation are undesirable too.   It is, nevertheless, possible that, in some cases as yet unknown, they may offer some advantages over other polymers.

Literature Cited

1. Goethals, E. J., *J. Macromol. Sci. (Reviews)* (1968) **C-2**, 73.
2. Zillberman, E. N., Teplyakov, N. M., *Vysokomolekul. Soed.* (1960) **2**, 133.
3. Coffman, D. D., McGrew, F. C., U.S. Patent 2,317,155 (1943).
4. Elias, H.-G., Bührer, H. G., *Makromol. Chem.* (1970) **140**, 21.
5. Schöberl, A., Krumey, F., *Ber. Dtsch. Chem. Ges.* (1944) **77**, 371.
6. Schöberl, A., Wiehler, G., *Angew. Chem.* (1954) **60**, 273.
7. Schöberl, A., Wiehler, G., *Liebigs Ann. Chem.* (1955) **595**, 101.
8. Schöberl, A., *Makromol. Chem.* (1960) **37**, 64.
9. Greber, G. (1970) private communication.
10. Frehsen, H., Ph.D. thesis, Hannover Technical University (West Germany), 1960.
11. Bührer, H. G., Elias H.-G., *Makromol. Chem.* (1970) **140**, 41.
12. Knunyants, I. L., Kuleshova, N. D., Lin'kova, M. G., *Izv. Akad. Nauk SSSR, Ser. Khim.* (1965) 1081; *C. A.* (1965) **63**, 8192 g.
13. Fleš, D., Markovac-Prpic, A., Tomasic, V., *J. Amer. Chem. Soc.* (1958) **80**, 4654.
14. Fleš, D. and Tomasic, V., *J. Polym. Sci.* (1968) **B-6**, 809.
15. Knunyants, I. L., Pervova, E. Ya., Lin'kova, M. G., Kil'disheva, O. V., *Khim. Nauka i Prom.* (1958) **3**, 278. C. A. 52, 19929a (1958).
16. Sweeny, W., Casey D. J., U.S. Patent 3,367,921 (1968).

17. Overberger, C. G., Weise, J. K., *J. Amer. Chem. Soc.* (1968) **90**, 3533.
18. Korte, F., Glet, W., *J. Polym. Sci.* (1966) **B-4**, 685.
19. Korte, F., Christoph, H., *Ber. Dtsch. Chem. Ges.* (1961) **94**, 1966.
20. Overberger, C. G., Weise, J. K., *Polym. Sci.* (1964) **B-2**, 329.
21. Overberger, C. G., Weise, J. K., *J. Amer. Chem. Soc.* (1968) **90**, 3538.
22. Marvel, C. S., Kraiman, E. A., *J. Org. Chem.* (1953) **18**, 1664.
23. Houben-Weyl, "Methoden der Organischen Chemie," Vol. 9, p. 745, G. Thieme, Stuttgart, West Germany (1955).
24. Janssen, M. J., *in* S. Patai, Ed., "The Chemistry of Carboxylic Acids and Esters," Interscience, New York (1969).
25. Reid, E. E., "Organic Chemistry of Bivalent Sulfur," Vol. 4, p. 24, Chemical Publishing, New York (1962).
26. Patrick, J. C., Ferguson, H. R., U.S. Patent **2,563,133** (1951).
27. Murphey, W. A., U.S. Patent **2,870,126** (1959).
28. Marvel, C. S., Kotch, A., *J. Amer. Chem. Soc.* (1951) **73**, 1100.
29. Flory, P. J., U.S. Patent **2,510,567** (1950).
30. Marvel, C. S., Kraiman, E. A., *J. Org. Chem.* (1953) **18**, 707.
31. Martin, J. C., Gilkey, R., U.S. Patent **3,254,061** (1966).
32. Stimpson, J. W., British Patent **853,730** (1960).
33. British Patent **783,546** (1957).
34. Korshak, V. V., Vinogradova, S. V., Lebedeva, A. S., U.S.S.R. Patent **203,902** (1967).
35. Noda, L. H., Kuby, St. A., Lardy, H. A., *J. Amer. Chem. Soc.* (1953) **75**, 913.
36. Schwyzer, R., *Helv. Chim. Acta* (1953) **36**, 414.
37. Nyquist, R. A., Potts, W. J., *Spectrochim. Acta* (1959) **7**, 514.
38. Sjöberg, B., *Z. Physik. Chem.* (1942) **52B**, 209.
39. Kuriyama, K., Komeno, T., Takeda, K., *Ann. Rep. Shionogi Res. Lab.* (1967) **17**, 72.
40. Schöberl, A., *Angew. Chem.* (1948) **A-60**, 7.

RECEIVED February 2, 1972.

Heat-Resistant Polyarylsulfone Exhibiting Improved Flow during Processing

ROBERT J. CORNELL

Uniroyal Chemical, Division of Uniroyal, Inc., Naugatuck, Conn. 06770

The reaction of bis(4-chlorophenyl)sulfone with the anhydrous dialkali salt of α,α'-bis(4-hydroxyphenyl)-p-diisopropyl benzene, which is dissolved in an aprotic solvent, leads to a new heat-resistant polyarylsulfone. A general discussion of monomer and polymer preparation is given as well as physical and mechanical properties of this polyarylsulfone. The polyarylsulfone thermoplastic exhibits essentially the same heat resistance of the polyarylsulfone based on bis-(4-chlorophenyl)sulfone and 2,2-bis(4-hydroxyphenyl)propane, along with the added plus of a lower melt viscosity at equivalent processing temperatures. The flow improvement is demonstrated by comparison with Brabender data, injection-molding conditions, and melt-viscosity data.

In recent years, there has been a considerable amount of interest in high-molecular-weight polyarylsulfone polymers (*1-3*). These polymeric materials possess heat-deflection temperatures of 165°-260°C plus excellent thermal stability at the high processing temperatures required. This paper deals with the synthesis and the physical and mechanical properties of a new heat-resistant polyarylsulfone. This new polyarylsulfone thermoplastic exhibits a significant reduction in apparent (melt) viscosity with only a slight reduction in heat deflection temperature when compared with the polyarylsulfone based on bis(4-chlorophenyl)sulfone and 2,2-bis(4-hydroxyphenyl)propane.

Experimental

Reagent. Tetrahydrothiophene 1,1-dioxide (commonly called sulfolane) was obtained from Shell Chemical containing 3% water. Anhydrous sulfolane was obtain by distillation under reduced pressure. Reagent grade potassium hydroxide, chloroform, and benzene were used without further purification.

Monomers. BIS(4-CHLOROPHENYL)SULFONE. We prepared bis(4-chlorophenyl)sulfone (I in Reaction 1) by dropwise addition of chlorosulfonic acid (0.4 mole) to chlorobenzene (0.2 mole) at 0°C. One hour after the chlorosulfonic acid addition was completed, an additional quantity of chlorobenzene (0.16 mole) was added to the cold solution.

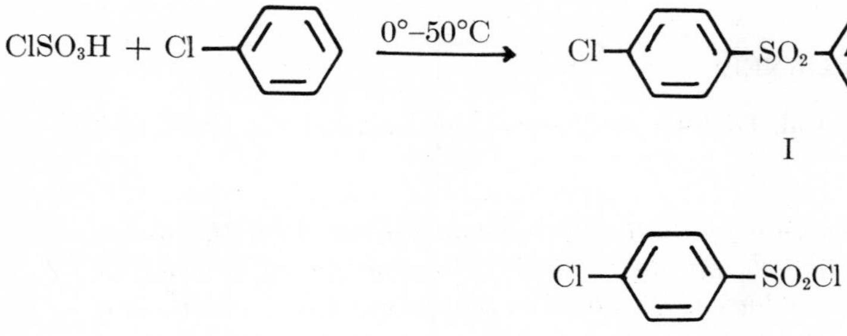

The reaction mixture was allowed to warm gradually to 50°C. The reaction was quenched by pouring into ice water. The aqueous suspension was heated to hydrolyze p-chlorobenzenesulfonyl chloride (II) to the water-soluble sulfonic acid. The desired product was filtered and washed with water until essentially neutral. Bis(4-chlorophenyl)sulfone was purified by recrystallization from benzene—m.p. 145°-147°C (4). Extremely low yields were obtained by this process. By contrast, yields in excess of 90% have been reported in processes that involve the reaction of chlorobenzene with sulfur trioxide and dimethyl or diethyl sulfate (5, 6).

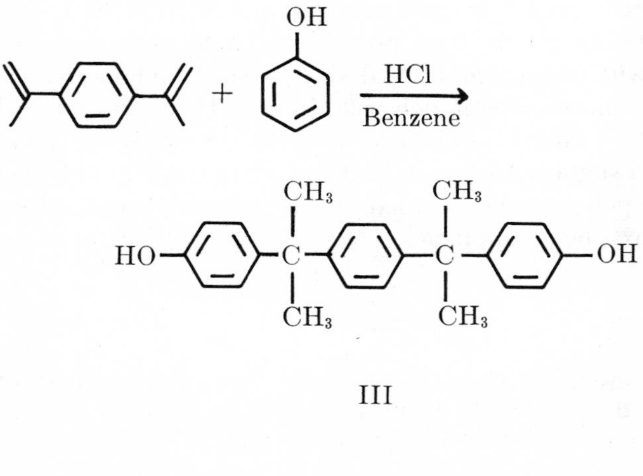

α,α'-BIS(4-HYDROXYPHENYL)-p-DIISOPROPYLBENZENE. α,α'-Bis(4-hydroxyphenyl)-p-diisopropylbenzene (III in Reaction 2) was prepared by dropwise addition of p-diisopropenylbenzene in toluene to a $3M$ excess of phenol saturated with gaseous hydrogen chloride. After a few hours, the bulk of the α,α'-bis(4-hydroxyphenyl)-p-diisopropylbenzene formed crystallized out. The excess phenol can be removed by steam stripping. The compound was purified by recrystallization from acetone—m.p. 191°-192°C (7). The bisphenol can also be prepared from α,α'-dihydroxy-p-diisopropylbenzene and phenol (8, 9).

Polymerization. Typically, the dihydric phenol (1 mole) and aqueous alkali metal hydroxide (2 moles) are mixed under an inert atmosphere in sulfolane and benzene. The water from the aqueous solution plus metal phenoxide formation is removed by distillation of a benzene-water azeotrope between 110° and 140°C. After water removal has been completed, the excess benzene is distilled off, the anhydrous salt in sulfolane cooled to 70°-80°C, and bis(4-chlorophenyl)-sulfone (I) added. The temperature is increased gradually to 200°C and held for four to five hours. Methyl chloride is bubbled in at the end of the polymerization to convert any terminal phenoxide groups to methyl ethers (10).

$$(3)$$

$$n = 15{-}30$$

The resin (IV) can be isolated from the mass by filtration of the alkali metal chloride and stripping of the solvent, or by precipitation in a nonsolvent. A second method involves dispersion of the polymer mass in water to remove the alkali metal halide, formed during the polymerization, and the sulfolane. Drying may be accomplished by heating the polyarylsulfone to 100°-120°C *in vacuo* for 8 to 12 hours.

The dried polyarylsulfone (IV) is soluble in halogenated hydrocarbons such as chloroform, chlorobenzene, and methylene chloride.

Molding. Test specimens were either compression molded in a hydraulic press at 245°C or injection molded at 275°C using a 2-ounce Ankerwerk, Model 75.

Physical Property Measurements. Flexural modulus and strength were measured according to ASTM Method D-790. Impact resistance was measured by the Izod notch ASTM D-256, and Rockwell Hardness according to ASTM D-785. Heat-distortion temperatures were determined by ASTM Method D-648 at 264 psi stress loading.

Mixing or torque generation data were obtained using a Brabender PlastiCorder. The Brabender PlastiCorder has a small mixing cavity containing two rotating mixing blades; the speed of the mixing blades may be varied. The mixing cavity has a jacket through which heated oil may be circulated. The polymer to be examined is added to the cavity. The rotating blades exert a torque that can be measured and depends on the viscosity of the polymer.

Apparent viscosity data were obtained on the Instron Capillary Rheometer, Model TT. For this purpose, rods 5 inches by $\frac{3}{8}$ inch were prepared by compression molding from material to be tested. Each rod was heated to 260° or 275°C in the barrel of the rheometer for five minutes. A piston plunger is then pressed down on top of the heated rod, forcing the rod to flow through a 0.060-inch diameter capillary, having a length-to-diameter ratio of 33. The piston plunger descends at a constant speed of from 0.005 to 5 inches per minute. The force required to extrude the rod through the capillary is measured.

Molecular-weight distributions of resins IV and V (*see* Reaction 4) were determined on an Anaprep gel permeation chromatograph. Four columns were used: 10^6A, 10^5A, 10^4A, and 10^3A. Two-milliliter samples of a 0.25% solution in o-dichlorobenzene at 100°C were injected. The solution was collected at a rate of 2 ml per minute.

Results and Discussion

Polymer Evaluation. Physical properties, melt-viscosity data, and gel permeation chromatography data of our polyarylsulfone IV and the polyarylsulfone V based on bis(4-chlorophenyl)sulfone and 2,2-bis(4-hydroxyphenyl)propane (*11*) are listed in Tables I through IV and Figures 1 through 3.

$$(4)$$

V

Resin IV has a lower specific gravity than does resin V (Table I), probably explained by the presence of two isopropylidene groups per mer unit. These reduce the ability of the polymeric chains to be as tightly packed as in resin V. The lower specific gravity of IV cannot be explained by the per cent crystallinity present in both polymers (Table I). If crystallinity were the major force, the specific gravity of resin IV would be higher. The remaining properties of resins IV and V listed in Table I exhibit little variation.

Table I.   Physical Properties of Polyarylsulfone Resins IV and V

Property	Resin IV	Resin V
Rockwell "R"	127	128
Specific Gravity	1.19	1.24
HDT 264 psi (°C)	158	166
Izod ¼″ (RT) (ft. lbs./in. notch)	0.8	1.1
Izod ¼″ (-29°C)	0.8	1.3
Izod ⅛″ (RT)	0.8	1.3
Izod ⅛″ (-29°C)	0.8	1.3
Tensile Strength (psi)	11,350	10,370
Tensile Modulus (psi) $\times 10^5$	3.7	3.5
Tensile Elongation (%) at break [a]	10	6.7
Flexural Strength (psi) $\times 10^5$	16,730	14,370
Flexural Modulus (psi)	4.0	3.4
Per cent Crystallinity	4.1	~0.0

[a] Samples annealed 120°C for 100 hours.

Table II.   Torque-Generation Data Obtained During Brabender Mixing of Resins IV and V

Material	RPM	Melt Temp. (°C)	Torque (gram-meters)	Mixing time (minutes)
Resin IV	50	227	2,300	4
	50	235	1,500	10
	60	235	1,800	8
Resin V	50	235	3,800	12
	60	250	2,600	15

Table III.   Injection-Molding Conditions for Polyarylsulfone Resins IV and V

	Resin IV	Resin V
Melt Temperature (°C):	290	290
Injection Cylinder Temp.		
Nozzle (°C):	275	275
Zones 1 and 2 (°C):	275, 275	275, 275
Injection Pressure		
T-bar (psi)	1300	1600
Shots (psi)	550	800
Mold Temperature (°C):	85	85
Back Pressure (psi)	200	200

The significant improvement in flow properties of resin IV *vs.* V is evident from the data. First, the torque from Brabender mixing indicates that resin IV is an easier-flowing material (Table II). The lower torque values for IV indicate the necessity of a lower-energy input to mix the polymer melt. This lower rotational force therefore indicates the polymer melt has a lower melt viscosity. Secondly, injection-molding conditions demonstrate the improved processability of resin IV (Table III) in comparison with resin V. At the same injection cylinder temperatures, the injection pressure for the tensile bar and Izod/heat distortion bar molds is lowered by 300 psi and 250 psi, respectively.

Finally, the apparent viscosity, at various shear rates, of resin IV at 500° and 525°F is lowered by a factor of 3 to 4 (Table IV and Figures 1 and 2). The apparent viscosity of resin IV at 500°F still is lower than resin V at 525°F.

Table IV.   Apparent Viscosity Data of Polyarylsulfone Resins IV and V

	Apparent Viscosity			
	(Dynes sec/cm²)			
	Resin IV		Resin V	
Shear rate (sec⁻¹)	260°C	175°C	260°C	275°C
0.173	—	—	35.3×10^4	—
0.433	8.70×10^4	—	34.7	16.8×10^4
0.865	9.81	2.71×10^4	31.5	15.2
1.73	9.25	3.54	30.0	13.3
4.33	8.08	3.90	28.2	11.8
8.65	7.78	3.81	27.8	10.8
17.3	7.10	3.67	28.3	10.1
43.3	6.50	3.25	48.8	8.90×10^4
86.5	5.67	3.17	—	7.39
173	5.06	2.40	—	5.71
433	—	1.44	—	—
865	—	0.96	—	—
1730	—	0.70	—	—

Figure 1, shows a rapid increase in apparent viscosity for resin V as the shear rate reaches 17.3 sec⁻¹. The sharp increase in apparent viscosity is not any chemical change, such as crosslinking of polymeric chains. The apparent viscosity curve can be retraced by lowering the shear rate. This increase in apparent viscosity can be eliminated by increasing the measurement temperature, as shown in Figure 2. This same phenomenon has been reported for polystyrene by Penwell and Porter (12). The explanation of the apparent viscosity increase in capillary flow of polystyrene was quantitatively explained through the pressure dependence of glass transition, T_g.   The same explanation could account for the increase in apparent viscosity of resin V.   As the capillary pressure increases with increasing shear rate, T_g approaches the measurement temperature, causing a rapid rise in viscosity.

The above data substantiate this flow improvement but do not explain its occurrence.   A possible reason for the improvement in flow

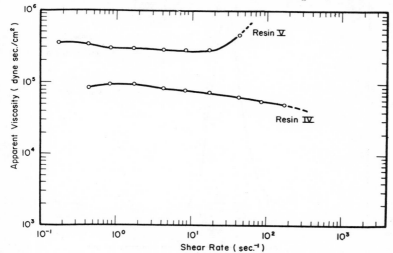

Figure 1.   Apparent viscosity vs. *shear rate of resins IV and V at 260°C*

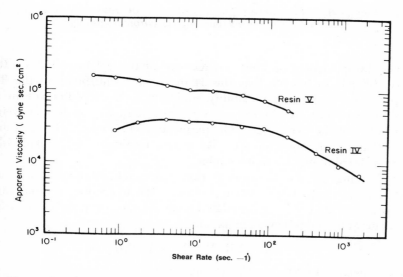

Figure 2.   Apparent viscosity vs. *shear rate of resins IV and V at 275°C*

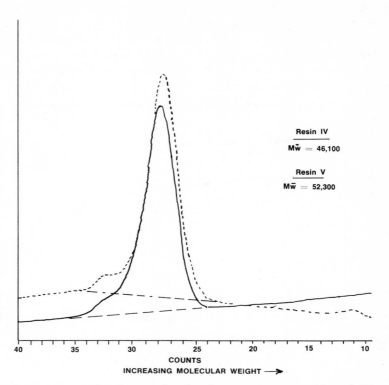

Figure 3. Molecular-weight distribution of resins IV and V determined by gel permeation chromatography

would be the looser packing arrangement of the polymer chains. It is very difficult to believe that this is the major contributor to the flow improvement. Another possibility could be a drastic difference in the molecular weight distribution or weight-average $(\overline{M}_w)$ molecular weight (or both) of the two resins. Gel permeation chromatography data indicate that this is not the case. As shown in Figure 3, the molecular-weight distributions are quite similar.

Working on the assumption that polystyrene standards are valid calibration standards for polyarylsulfones, the molecular weights $(\overline{M}_w)$ for resins IV and V were calculated from the molecular-weight distribution curves by the following equation.

$$M_w = \Sigma W_i M_i; \ W_i = H_i / \Sigma H_i$$

H_i = height of curve at various counts
ΣH_i = total of all heights measured
M_i = molecular weight of polystyrene at various counts.

Even if the assumption that polystyrene standards can be used is not completely valid, the error in the determined $\overline{M}_w$ molecular weights would be a constant, and should not alter the percent difference in M_w molecular weight of the two polyarylsulfones. The $\overline{M}_w$ molecular weight for resins IV and V are 46,100 and 52,300, respectively. This 10% difference could be significant if resins IV were to lie below the critical molecular weight (M_b) for polymer chain entanglement and resin V lie above. Much experimental data are available for the viscosity of polymer melts. These data, without exception, obey this equation for M greater than some critical molecular weight M_b:

$$\eta = KM^{3.5}$$

This relationship is shown in Figure 4 for polystyrene (*13*).

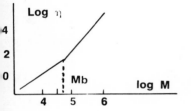

Figure 4. Variation of melt viscosity of polystyrene as a function of molecular weight at 217°C

The line for M greater than M_b has a slope of 3.5. If resins IV and V do, in fact, lie on opposite sides of the critical molecular weight (M_b), a 10% increase in $\overline{M}_w$ molecular weight could be accompanied by a 40% or higher increase in melt viscosity. To substantiate this possible explanation for the significant differences in melt viscosity of resins IV and V, M_b for both resins would have to be determined.

Acknowledgment

The author wishes to thank the entire Polymer Physics Group for helpful assistance on polymer characterization, Peter M. Byra for handling the injection-molding operations, and Angenette Grant for her assistance in the monomer and polymer syntheses.

Literature Cited

1. Lee, H., Stoffy, D., Neville, K., "New Linear Polymers," McGraw-Hill, New York (1967).
2. Rose, J. B., Jennings, B. E., Jones, M. E. B., *J. Polym. Sci., Part C* (1967) 715.
3. Vogel, H. A., *J. of Polym. Sci., Part A-1* (1970) 8, 2035.

4. Erickson, A., U.S. Patent **2,860,168** (1958).
5. Joyl, R., Bucourt, R., Fabignon, C., U.S. Patent **2,971,985** (1961).
6. Keogh, M. J., Ingberman, A. K., U.S. Patent **3,415,887** (1968).
7. Ruppert, H., Schnell, H., British Patent **935,061** (1963).
8. Simmons, P., Ulyatt, J. M., British Patent **932,811** (1963).
9. Broderick, G. F., Oxenrider, B. C., Vitrone, J., U. S. Patent **3,393,244** (1968).
10. Cornell, R. J., U.S. Patent **3,554,972** (1971).
11. Johnson, R. N., Farnham, A. G., Clendinning, R. A., Hale, W. F., Merrian, C. N., *J. Polym. Sci., Part A-1*, (1967) **5**, 2375.
12. Penwell, R. C., Porter, R. S., *Appl. Polym. Sci.*, (1969) **13**, 2427.
13. Bueche, F., "Physical Properties of Polymers," Interscience, New York (1962).

RECEIVED April 10, 1972.

Polymerization by Opening of Small Carbon Rings

C. P. PINAZZI, J. C. BROSSE, A. PLEURDEAU, J. BROSSAS, G. LEGEAY, and J. CATTIAUX

Laboratoire de Chimie Organique Macromoléculaire, Equipe de Recherche Associée au Centre National de la Recherche Scientifique, Rte de Laval 72, Le Mans, France

Polymerization of cyclopropane and substituted cyclopropyl compounds was studied with cationic initiators and Ziegler-Natta catalysts. Cyclopropane, bicyclo[n.1.0]alkanes, spiro [2.n]alkanes, bicyclopropyle, and two isomers of isoprene give rise, with Lewis acids as initiators, to oligomers whose structures generally present methyl groups in a side chain. These methyl groups stem from molecular rearrangement involving opening of the cyclopropyl group and hydride shift during the polymerization step. With Ziegler-Natta catalysts, these monomers give oligomers whose structures are different from those observed with cationic catalysts. Dihalocyclopropyl compounds give, with either cationic or Ziegler-Natta catalysts, oligomers by opening of the three-carbon ring. The structures of these polymers are the same in both cases and are characterized by the loss of one molecule of HCl per monomer unit.

This work concerns the study of the polymerization of cyclopropane, substituted cyclopropanes, and conjugated cyclopropanes in the presence of cationic and Ziegler-Natta polymerization. The unsaturation of cyclopropane has been described by several workers in the same way as unsaturated compounds. The unsaturation of cyclopropane compounds, which is the basis for the polymerization of these structures, can be explained by the electronic repartition on the three carbon atoms of the ring. Determination of the dipolar moment of chlorocyclopropane has shown that the carbonium ion resulting from the attack of the ring by a carbo cation is stabilized in a homoallylic structure.

Several hypotheses concerning the electronic structure of cyclopropane have been suggested. This topic has been interestingly pinpointed by Bernett (1). A cyclopropane model proposed by Walsh (2) indicates that C–C bonds of the rings are caused by an overlap of one of the sp^2 hybridized orbitals of each carbon atom and of each orbital (Walsh proposed, in fact, an sp^2 hybridization state for each cyclopropane carbon).

Consideration of the unsaturation of cyclopropane led the author to compare the attack of a proton on cyclopropane with the formation of a π complex. Hückel (3) determined mathematically the most stable structure for a protonated cyclopropane. The representation given is imperfect since it is concerned with only one resonant structure; there are three possibilities as one of the p orbitals is antibonding. In fact, it has been calculated that the bonds formed by the overlap of the p orbitals are occupied by four electrons, while only two electrons are associated in the overlap of three sp^2 orbitals in the center of the ring. In this representation, the C–H bonds are represented by sp^2 hybrid orbitals.

A more recent representation of cyclopropane has been proposed by Coulson and Moffitt (4), who introduced the notion of a bent bond. This representation has the advantage of minimizing the bond energies. The orbitals associated with the C–C bonds were calculated as being $sp^{4,12}$ hybridized, while those of the C–H bonds are $sp^{2,28}$ hybridized (5). Under these conditions, the valence angle is 104°, which corresponds to a bending of 22°. Coulson and Goodwin reconsidered this problem, applying the principle of maximum orbital overlap (6). The bending then assumes a value of 21°26'. Finally, the work of Randic and Maksic (7) gives a value of 101°32' for the C–C valence angle, which involves a bending of 20°46'. The optimum conformation is then obtained when the C–C bond orbitals are sp^5 hybridized and the C–H orbitals sp^2 hybridized.

The unsaturated nature of cyclopropane and its derivatives suggests that they are able to polymerize with participation of the ring the same as C=C compounds. The first studies were made by Tipper and Walker (8), who used a cationic catalyst ($AlBr_3$–HBr) at lower temperature (between 0° and −78°). They showed that the mechanism was similar to that of the polymerization of propylene and other olefins with Friedel-Crafts catalysts.

A more recent paper (9) indicates the existence of methyl groups in side chains in cationic polymers of cyclopropane. This is not in conflict with the existence of a π complex during the initiation step, but it suggests that the polymerization mechanism is more complex than that proposed by Tipper and Walker.

Ketley (*10*) has shown that 1,1-dimethylcyclopropane gives the same polymer by cationic polymerization as that obtained from 3-methyl-1-butene. He assumed a π complex mechanism for initiation.

Isopropylcyclopropane treated with Lewis acids polymerizes by participation of the cyclopropyl group (*11*). In this case no evidence has been found for a hydride shift occurring in the 3-methyl-1-butene polymerization mechanism. Phenylcyclopropane has been polymerized with Lewis-acid-type initiators (*12*) and with Ziegler-Natta catalysts (*13*). The polymers have a structure that implies opening of the three-carbon ring. Norcarane (bicyclo [4.1.0] heptane) reacts with Ziegler-Natta catalysts (*13*) and gives oligomers having cyclohexane units in the chain. It polymerizes by opening of the small ring, leaving the cyclohexane unchanged.

This paper deals with the polymerization of several purely hydrocarbonated and *gem*-dihalocyclopropanic monomers. The influence of conjugation between a cyclopropane and one double bond, and the influence of the conjugation between two cyclopropanic groups is shown for several monomers, such as spiropentane, vinyl *gem*-dihalocyclopropane, and bicyclononene.

Polymerization of Cyclopropanic Systems

Tipper and Walker (*8*) have already studied the kinetics of the polymerization of cyclopropane and cyclobutane in heptane between $0°$ and $-78°C$, using a $AlBr_3$–HBr catalyst system. We have studied the structural aspects by subjecting cyclopropane (M_1) and cyclobutane to different types of initiators. Using a cationic process in the presence of $SnCl_4$, $TiCl_4$, Et_2O–BF_3, and $AlBr_3$ between $-30°$ and $100°$ polymerization of cyclopropane occurs, whereas the same initiators have no effect on cyclobutane. The mechanism involves the formation of a π complex with one of the cyclopropane bonds followed by a hydride shift, giving rise to structure P_1. NMR and IR spectroscopy show that the polymer has a polypropylene structure.

In the presence of Ziegler-Natta catalysts, in the heterogeneous phase, the reactivities of cyclopropane and cyclobutane are similar. The conversion degrees vary between 1 and 5%, and the molecular

M_1 P_1 P_2 (1)

weights are $1500 < \overline{M}_n < 2000$ for the soluble fraction of the polymers. With this kind of catalyst, the polymers have a polyethylene structure of type P_2 (Equation 1).

Polymerization of Bicyclo[n.1.0]alkanes and Spiro[2.n]alkanes. GENERAL PREPARATION. The monomers were synthesized by addition of methylene to the double bond of the corresponding cycloalkenes or methylenecycloalkenes—Simmons and Smith reaction (14)—or by addiition of a dihalocarbene followed by the reduction of the dihalocyclopropane group by a Na/hydrated methanol system (Equation 2) (15, 16):

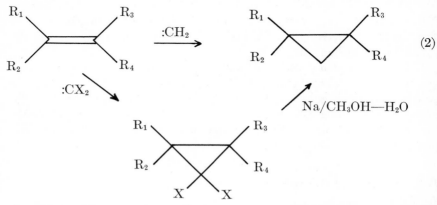

(2)

In the first case, the methylene is obtained by reaction of a Zn-Cu complex on diodomethane. The addition reaction itself is carried out in the heterogeneous phase. The yields in bicyclo compounds depend on the condition of addition to the cycloalkene double bond. The interest in this method lies in the fact that the starting cycloolefins are commercially available and that the yields in bicycloalkanes are good (from 30 to 80%). A disadvantage, however, is that the final products are relatively hard to separate, especially the high-molecular-weight monomers.

POLYMERIZATION OF BICYCLO[n.1.0]ALKANES. From the large-ring bicyclo[n.1.0]alkane series, six were chosen because of their relative ease of synthesis: bicyclo[5.1.0]octane (M_2), bicyclo[6.1.0]nonane (M_3), bicyclo[10.1.0]tridecane (M_4), and the corresponding 1-methylbicyclo-[n.1.0]alkanes and prepared from the corresponding cycloalkenes or 1-methylcycloalkenes (cycloheptene, cyclooctene, and cyclododecene). See Equation 4.

CATIONIC POLYMERIZATION OF BICYCLO[n.1.0]ALKANES. The bicyclo [n.1.0]alkanes were polymerized in methylene chloride by various Lewis acids: $TiCl_4$, BF_3–Et_2O, and $SnCl_4$—with catalyst/monomer molar ratios from 4 to 15%. The temperatures used were between 20° and 80°C. At the lower temperatures and for the concentrations used, the degree

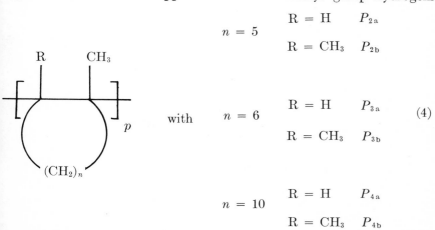

$$n = 5 \qquad M_2 \quad \begin{array}{l} (\text{R} = \text{H}) \\ (\text{R} = \text{CH}_3) \end{array}$$

$$n = 6 \qquad M_3 \quad \begin{array}{l} (\text{R} = \text{H}) \\ (\text{R} = \text{CH}_3) \end{array} \qquad (3)$$

$$n = 10 \qquad M_4 \quad \begin{array}{l} (\text{R} = \text{H}) \\ (\text{R} = \text{CH}_3) \end{array}$$

of conversion to polymers is very low, suggesting that polymerization of these compounds necessitates relatively high temperatures.   The degrees of polymerization are low and vary little with the experimental conditions, and they decrease as the ring size increases.   Steric hindrance effects are therefore probably responsible for oligomer production. Conversions vary from 0 to 65%, and are lower for bicyclo[10.1.0]tridecane than for the other two monomers.   For a given monomer, the degree of conversion depends on the amount of catalyst.

The polymer structures were studied by infrared spectroscopy and NMR.   Infrared is ineffective in distinguishing between polymers of these three types.   In all three cases, C–H bonds signals are observed at 2950, 2920, and 1450 cm^{-1}, and a band at 1375 cm^{-1} may be attributed to methyl groups.

NMR spectroscopy gives more detail and distinguishes between the three types of polymers.   The signals recorded in the region of $\delta = 1.4$ ppm are caused by hydrogens carried by the ring carbons, whereas the signals situated at 0.8-1 ppm are those of methyl group hydrogens

$$n = 5 \qquad \begin{array}{ll} \text{R} = \text{H} & P_{2a} \\ \text{R} = \text{CH}_3 & P_{2b} \end{array}$$

with $\quad n = 6 \qquad \begin{array}{ll} \text{R} = \text{H} & P_{3a} \\ \text{R} = \text{CH}_3 & P_{3b} \end{array} \qquad (4)$

$$n = 10 \qquad \begin{array}{ll} \text{R} = \text{H} & P_{4a} \\ \text{R} = \text{CH}_3 & P_{4b} \end{array}$$

carried by the saturated carbons. It must therefore be assumed that polymerization proceeds by opening of the cyclopropane, followed by a rearrangement leading to the formation of one methyl group per "monomer unit" and two in the case of 1-methylbicyclo[n.1.0]alkanes; *see* structures P_{2a} to P_{4b} (Equation 4).

Of various mechanisms that may be proposed, the only acceptable one is that summarized in Equation 5. It is assumed that the attack on the cyclopropane system by the active site leads to the formation of a π complex, which later rearranges to a carbo cation. The rupture of the bond of carbons 1 and 2 and the rotation between A and carbon 2 involves the appearance of a positive charge on carbon 1. The primary carbo cation formed will be able to rearrange into a more stable tertiary carbo cation by hydride shift. The polymers obtained by such a mechanism would have structures P_{2a} to P_{4b}. They are the only ones having one methyl group in the side chain per "monomer unit" and two in the case of 1-methylbicyclo[n.1.0]alkanes. It must, therefore, be assumed that this is the mechanism to be considered, and that structures P_{2a} to P_{4b} are the only ones that agree with the data.

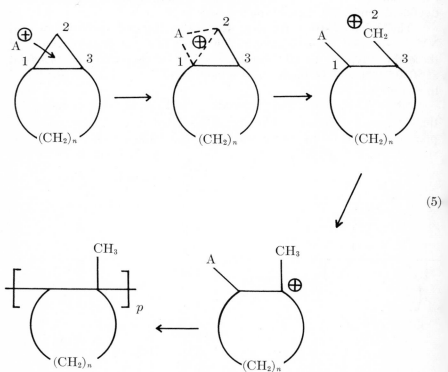

(5)

The initiation step may be interpreted by assuming that a proton gives rise to a π complex with the cyclopropane, which rearranges by hydride shift into a carbo cation that will be the active site for the polymerization. The termination step may be considered as a deprotonization in the α position from the active site of the last unit in the chain. This hypothesis is confirmed experimentally in oligomers having a very low degree of polymerization by the presence of weak NMR signals at $\delta = 5.2$ ppm (characteristic of vinylic protons) and at about $\delta = 2$ ppm (methyl groups carried by C=C).

These results agree with those of various authors who have demonstrated the transformation of bicyclo[n.1.0]alkane type structures under acid catalysis into compounds having a methyl group. However, it is not possible to visualize a prior isomerization into methylcycloalkenes followed by its polymerization. One must consider that polymerization occurs in a single step by transformation into the carbo cation. Even so, the presence of monomer units with the methyl group not in the side chain but carried by carbon atoms in the ring should not be excluded. If such units existed, they would be present in a very small proportion with respect to the main monomer unit.

POLYMERIZATION OF BICYCLO[n.1.0]ALKANES BY METAL-HALIDE COMPLEXES. Bicyclo[4.1.0]heptane (norcarane) has already been polymerized with Ziegler-Natta type catalysts (13). We have continued the study of the reactivities of the higher homologs of norcarane toward transition metal complex catalysts: bicyclo[5.1.0]octane, bicyclo[6.1.0]-nonane, and bicyclo[10.1.0]tridecane, and the corresponding 1-methyl-bicyclo[n.1.0]alkane (M_2 to M_4). The reactions were carried out in sealed tubes in hexane, under nitrogen, at 80°C for 24 hours. No reaction seems to occur at low temperatures. Various catalyst systems were used: titanium, vanadium, and tungsten halides used in conjunction with trialkylaluminum (Et_3Al, (i-Bu)$_3$Al, and Bu_3Al). The best conversions were obtained with the complexes of triethylaluminum, triisobutylaluminum, or tributylaluminum with stannic chloride. Complexes formed with titanium tetrachloride produce no polymer. Although oligomers are obtained in all cases, the average degrees of polymerization are generally higher than with cationic polymerizations (from five to eight on average), and are nearly identical for all six monomers.

Structural determination of the polymers obtained was difficult because of the complexity of the spectra. Infrared spectra show bands already observed in cationic polymerization at 2950, 2920, and 1450 cm^{-1}. The methyl band at 1375 cm^{-1} still exists, but its intensity is less. There is a peak at 1460 cm^{-1}, in the $-CH_2-$ region, which suggests that the polymers possess methylene groups different from the cyclic methylenes. NMR shows a signal at about $\delta = 1.4$ ppm, which may be attributed to ring hydrogens, and a signal at $\delta = 0.8\text{--}0.9$ ppm, characteristic of methyl groups on saturated carbons. The proportion of methyls is far less than

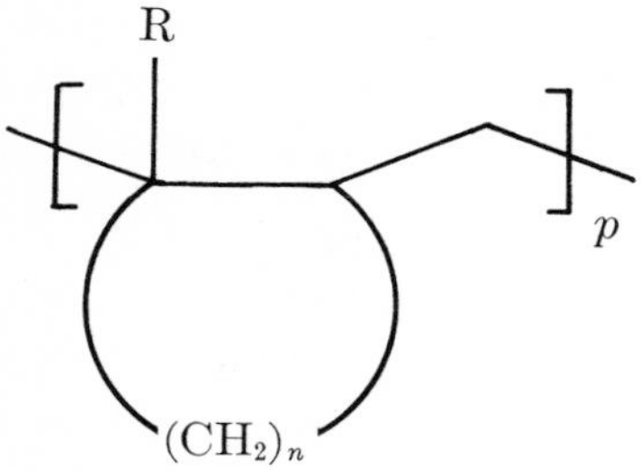

$$
\begin{array}{lll}
 & R = H & P_{2c} \\
n = 5 & & \\
 & R = CH_3 & P_{2d} \\
\\
 & R = H & P_{3c} \\
n = 6 & & \\
 & R = CH_3 & P_{3d} \\
\\
 & R = H & P_{4c} \\
n = 10 & & \\
 & R = CH_3 & P_{4d}
\end{array}
\tag{6}
$$

one per "monomer unit," and less than two with polymers of the 1-methylbicyclo[n.1.0]alkanes group. When it is assumed that the chain ends are ethyl, butyl, or isobutyl residues (depending on the type of catalyst), the monomer unit consists of a 7-, 8-, or 12-carbon ring, depending on the monomers considered, and that these rings are connected by a methylene group P_{2c} to P_{4d}. In particular, for catalysts using (i-Bu)$_3$Al, the CH_3/CH_2 ratio in the polymer produced is greater than for the catalysts obtained from Et$_3$Al. For polymers of higher degrees of polymerization, the methyl peaks are negligible, thus confirming that they belong to chain-end methyl groups. This hypothesis is confirmed by the presence of an NMR peak in the region of $\delta = 1.2$ ppm that does not appear in bicycloalkane cationic polymers (Equation 6).

The polymerization mechanism proposed is based on studies of olefin polymerization by Natta and Danusso (*17*). It involves attack of the cyclopropane by the complex $A_mMe\text{-}A'_1A'_2Me'$, where $Me = Ti$, Sn, V, or W; $Me' = Al$; $A = Cl$, or A'_1; and $A'_2 = Et$, i Bu, or *sec*-Bu (Equation 7).

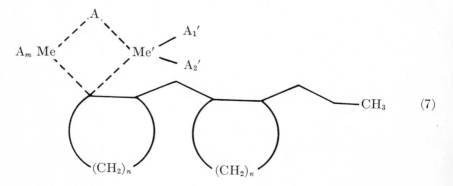

Polymerization of Spiro[2.*n*]alkanes.   Spiro[2.6]nonane (M_5), spiro [2.7]decane (M_6), and spiro[2.11]tetradecane (M_7) were prepared from the corresponding methylenecycloalkanes (methylenecycloheptane, methylenecyclooctane, and methylenecyclododecane, respectively) by the Simmons and Smith reaction (*14*); *see* Equation 8.

$$n = .6 \quad M_5$$

$$n = 7 \quad M_6 \tag{8}$$

$$n = 11 \quad M_7$$

The yield varies from 40 to 60% no matter what size the ring is. The separation of the alkenes and their derivatives is difficult because of the closeness of their boiling points. Furthermore, one obtains, probably as the result of partial isomerization during reaction of the methylenecycloalkanes to 1-methylcycloalkenes, small amounts of secondary compounds, such as the corresponding methylbicyclo[*n*.1.0] alkanes.

CATIONIC POLYMERIZATION OF SPIRO[2.*n*]ALKANES.   Cationic polymerization of spiro[2.5]octane and spiro[2.4]heptane has already been considered by Ketley and Ehrig (*18*) to compare the structures of the polymers obtained by means of AlBr$_3$ with those of polymers obtained from the corresponding vinylcycloalkanes.   These authors have found that the two groups of polymers have very different structures and, without giving definitive results, concluded that polymerization of spiranes probably occurs by passing through a bicyclic intermediate; on opening, the intermediate gives complex compounds.

Spiro[2.6]nonane, spiro[2.7]decane, and spiro[2.11]tetradecane have been polymerized in methylene chloride in the presence of Friedel-Crafts catalysts with this increasing order of reactivity: VOCl$_3$, SnCl$_4$, BF$_3$–Et$_2$O, TiCl$_4$, and AlCl$_3$.   The reactions were carried out over 24 hours at 80°C.   Polymerization, although requiring rather high temperatures (above 20°C) could occur at temperatures lower than those observed for the corresponding bicyclo[*n*.1.0]alkanes.   Conversions are generally higher than for the corresponding bicycloalkanes.   The polymerization degrees are low and virtually identical for the same operating conditions (from 3 to 6).

The infrared spectra of polymers show bands at 2950, 2920, and 1450 cm^{-1}.   The presence of a peak at 1375 cm^{-1} means that these polymers have a methyl group, which confirms the results of Ketley and Ehrig and proves that the structure is different from that of polymers of vinyl cycloalkanes. NMR gives two massive peaks, one in the region of $\delta = 1.4$ ppm, corresponding to hydrogens in the rings, and the other in the region of $\delta = 0.9$ ppm, corresponding to the hydrogens of a methyl group on a saturated carbon. Integration shows one methyl group per monomer unit (Equation 9).

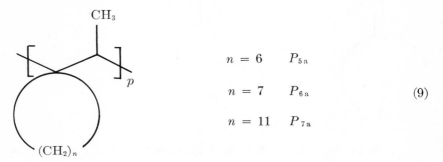

$$n = 6 \qquad P_{5a}$$
$$n = 7 \qquad P_{6a} \qquad (9)$$
$$n = 11 \qquad P_{7a}$$

The proposed polymerization mechanism suggests that a π complex is formed from the spirane cyclopropane and the carbo cation active site (Equation 10).

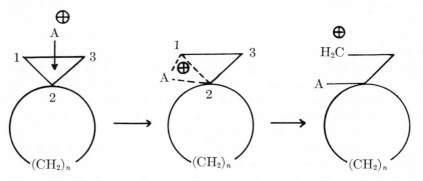

$$(10)$$

From this complex, the rupture of the bond between carbons 1 and 2 gives rise to the formation of a positive charge appearing on carbon 2, which involves formation of a primary carbo cation that can rearrange, by hydride shift, into a more stable secondary carbo cation on carbon 3, leading to polymers of structure P_{5a}, P_{6a}, and P_{7a}. These structures are the only ones having one methyl in the side chain per monomer unit.

The structure of cationic polymers of spiro[2.n]alkanes is too hindered to allow the formation of high-molecular-weight polymer. The termination step occurs rapidly, probably by rearrangement, loss of a proton, and formation of a double bond, the existence of which is confirmed by a vinyl hydrogen NMR signal at about $\delta = 5.2$ ppm.

POLYMERIZATION OF SPIRO[2.n]ALKANES BY TRANSITION-METAL COMPLEXES. Polymerization of spiro[2.n]alkanes by transition-metal complexes was carried out in hexane at 80°C over 24 hours. The catalyst systems were: $Et_3Al/TiCl_4$, $Et_3Al/SnCl_4$, $Et_3Al/VOCl_3$, and Et_3Al/WCl_6 for catalyst/monomer ratios of about 16%. The highest conversions (80 to 90%) were obtained with $Et_3Al/SnCl_4$, the highest polymerization degrees being observed with $Et_3Al/TiCl_4$ couple ($\overline{D}_p = 3$). This catalyst is practically without effect on bicyclo[n.1.0]alkanes and on methylbicyclo-[n.1.0]alkanes. It is possible that the more open structure of cyclopropane in the case of spirane-type compounds allows an easier reaction with this catalyst.

$$ n = 6 \qquad P_{5b} $$
$$ n = 7 \qquad P_{6b} \qquad\qquad (11) $$
$$ n = 11 \qquad P_{7b} $$

Infrared spectroscopy indicates bands at 2950, 2920, 1450, and 1375 cm^{-1} already observed on cationic polymers. However, the band at 1375 cm^{-1}, corresponding to methyl groups, decreases in intensity and completely disappears in high-molecular-weight polymers. Furthermore, the band at 1455 cm^{-1} may be attributed to linear $-CH_2-$ groups. NMR confirms this assumption, since the hydrogen in the rings appear in the region of $\delta = 1.4$ ppm, whereas the signal observed at $\delta = 0.9$ ppm obtained with cationic polymers is practically nonexistent. It appears at low intensity with low-molecular-weight oligomers, and can presumably be attributed to the chain-end methyl groups. From these results, structures P_{5b}, P_{6b}, and P_{7b} may be assigned to compounds obtained from spiroalkanes for this type of polymerization (Equation 11).

It is possible to assume a polymerization mechanism based on that proposed by Natta and Danusso (Equation 12) (*17*):

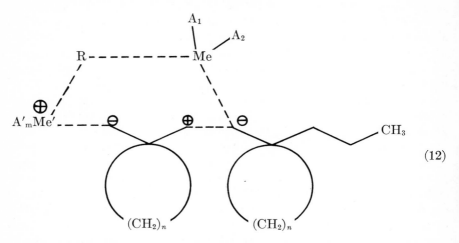

$$(12)$$

Polymerization of Spiropentane, Methylenecylobutane, and Bicyclopropyle. One compound in the spirane hydrocarbon series is especially worthy of attention—namely spiropentane (M_8), which is an isomer of both isoprene and methylenecyclobutane (M_9); *see* Equation 13. From a structural point of view, if one considers that spiropentane is the linear combination of the p orbital and the sp orbital associated with each ring of spiropentane, four equivalent sp^3 hybrid orbitals may be formed. Under these conditions, spiropentane constitutes a highly p-type unsaturated entity that is thus especially suitable for polymerization.

A slightly different monomer has also been studied: bicyclopropyle, composed of two cyclopropanes linked by a σ C–C bond (M_{10}). The conjugation between the two cyclopropanes of the monomer is similar to that appearing between the butadiene orbitals.

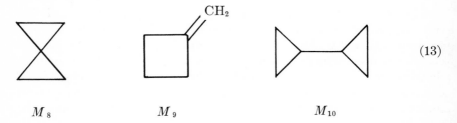

$$(13)$$

M_8 M_9 M_{10}

Syntheses of spiropentane, methylenecyclobutane, and bicyclopropyle are described in the appropriate sections.

POLYMERIZATION OF SPIROPENTANE. The preparation of spiropentane (*19*) is important, since other compounds are formed (for example, methylenecyclobutane, 2-methyl-1-butene) that are sensitive to cationic catalysis.   Spiropentane is synthesized from pentaerythrityl tetrabromide, which itself is prepared by successive reactions of hydrobromic acid and phosphorus tribromide on commerical pentaerythritol.   Pentaerythrityl tetrabromide is treated with zinc in alcoholic medium in presence of sodium ethylenediamine tetracetate (Equation 14).

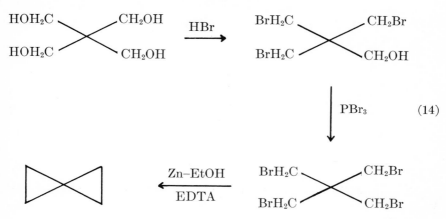

$$(14)$$

The residual alkenes are destroyed by bromine in dibromomethane, which leaves pure spiropentane (yield 60%), the purity being confirmed by a single NMR peak at 0.72 ppm.

CATIONIC POLYMERIZATION OF SPIROPENTANE.   These initiators, in decreasing order of activity—$AlCl_3$, WCl_6, $MoCl_5$, $ZrCl_4$, BF_3–Et_2O, $TiCl_4$, and $SnCl_4$—yield polymers for which the conversion degree is higher in methylene chloride than in *n*-hexane.   The temperature factor is important, since no reaction occurred below 20°C.   All runs have been carried out at 80°C or, occasionally, at 100°C.   The amount of initiator required is very high, the catalyst/monomer ratio being up to 10%.

The polymers obtained have an average molecular weight of about 1000, their infrared spectra having bands identical to those described for cyclized poly-1,4-isoprenes.   The peaks at 2960, 2925, and 2870 cm^{-1} correspond to cyclic methylenes.   A weak band lying between 1650 and 1700 cm^{-1}, the intensity of which depends on the cyclization degree, determines the proportion of tetrasubstituted cyclic double bonds (P_{8a}).   An absorption band between 1450 and 1465 cm^{-1} is caused by –CH_2– groups, and a peak at 1378 cm^{-1} characterizes the methyl groups.

Attempts of cyclization of high-molecular-weight poly-*cis*-1,4-isoprene has already been published (*20-22*).   The main characteristics of these compounds are (NMR) no signal for protons carried by a C=C double bond, but a strong signal at $\delta = 0.9$ ppm corresponding to protons from methyl groups on saturated carbons, and a massive peak between

$\delta = 1.1$ and $\delta = 1.7$ ppm, with maxima at $\delta = 1.25$ and 1.6 ppm, caused, respectively, by ring protons and protons of methyl groups on $C=C$ double bonds (structure P_{8a}). The relative size of these two maxima is a function of the cyclization degree, the peak at $\delta = 1.6$ ppm becoming smaller as the number of rings increases. However, it is difficult to determine this number exactly.

Cyclopolymers of spiropentane have the same characteristics as those of cyclopolyisoprenes. We conclude they have similar structures (P_{8a}) and (P_{8b}) in Equation 15.

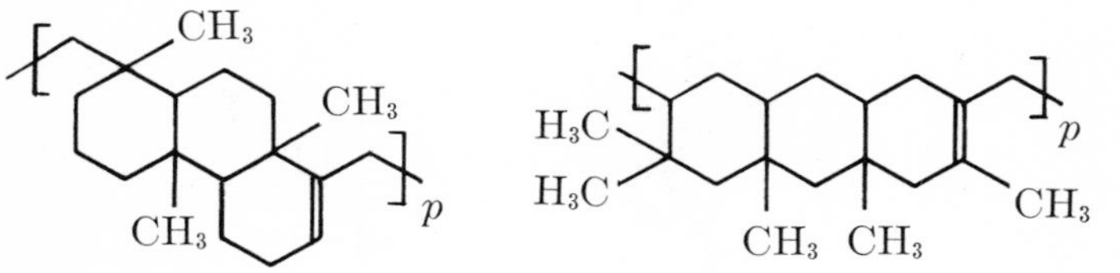

$$P_{8a} \qquad\qquad P_{8b} \tag{15}$$

The mechanism proposed involves the formation of a complex (15_1, Equation 15a) between the cationic initiator and the hydrocarbon ring, followed by its transformation into a carbo cation (15_2) similar to the spiropentylium ion proposed by Fanta (23). The intermediate structure (15_3) rearranges by cyclization because of the presence of Lewis acid and the rather high reaction temperature.

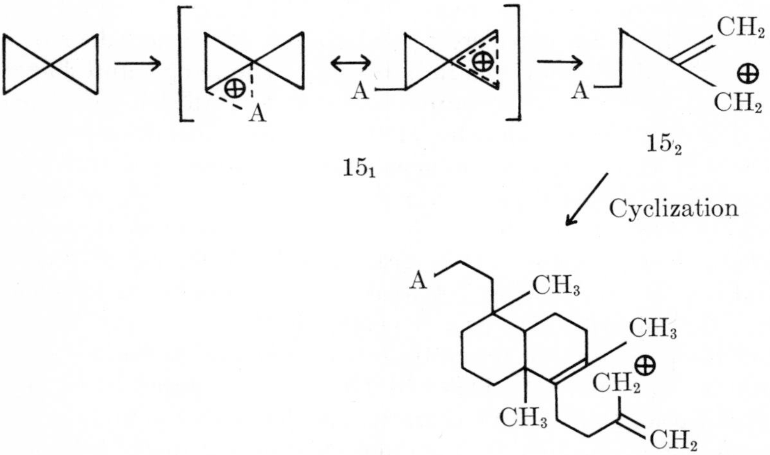

$$15_1 \qquad\qquad 15_2 \tag{15a}$$

Cyclization

$$15_3$$

POLYMERIZATION OF SPIROPENTANE BY METAL-HALIDE COMPLEXES.
Heterogeneous-phase polymerization with trialkylaluminum metal-halide

complexes has also been studied, using *n*-hexane as solvent at temperatures between 20° and 100°C (there is no reaction below 20°C). These catalysts were used: $TiCl_4/R_3Al$, WCl_6/R_3Al, $SnCl_4/R_3Al$, and $VOCl_3/R_3Al$, where R = Et, *i*-Bu, or Cl. As before, the polymers obtained are soluble in hydrocarbons and halogenated solvents except those produced by the reaction with the $SnCl_4/R_3Al$ couple, which gives insoluble products with conversion degrees of about 70%. The average molecular weights were determined by osmometry. They vary between 1000 and 5000, depending on the nature and amount of catalyst and temperature.

A structural study confirms cyclopolymerization. Infrared bands are obtained at 2960, 2925, and 2870 cm^{-1}, as with polymers obtained by Lewis acids. The difference between this case and the previous one lies in the existence of a shoulder of varying intensity on the methyl band between 1360 and 1370 cm^{-1} (the main peak is at 1380 cm^{-1}). This splitting corresponds to the presence of *gem*-dimethyl groups, the number of which varies inversely with the number of consecutive fused rings (structure P_{8b}). It is impossible to differentiate by NMR between cyclopoly-1,4-isoprenes and the 3,4 variety, since the signals obtained in both cases are identical.

In summary, cationic polymerization gives polymers of structure P_{8a}; Ziegler-Natta complexes mainly lead to blocks of structure P_{8b}. This study gives further evidence for cyclization reactions of high-molecular-weight polyisoprenes. This point of view has been confirmed by the study of the cyclization of model polyisoprene molecules with two, three, or four monomer units toward the catalysts able to initiate such a reaction (*24*).

POLYMERIZATION OF METHYLENECYCLOBUTANE. Methylenecyclobutane was synthesized by reaction of a Zn–Cu couple on pentaerytrityl tetrabromide (Equation 16). The yield of the synthesis is about 65% (*9*).

$$BrH_2C \diagdown \diagup CH_2Br \qquad BrH_2C \diagup \diagdown CH_2Br \xrightarrow{\;Zn/Cu\;} \qquad \qquad (16)$$

$$M_9$$

CATIONIC POLYMERIZATION OF METHYLENECYCLOBUTANE. Attempts were made to discover catalyst systems able to attack methylenecyclobutane, and tests were carried out on different classes of catalysts. There was no detectable reaction when using either benzoyl peroxide, or anionic catalysts such as sodium naphthalene. However, interesting results were obtained with Lewis acids, Ziegler-Natta catalysts, and complexes such as transition-metal acetyl acetonates/alkylaluminum.

In ionic polymerization, Equation 17, it can be assumed that there is a first stage comprising formation of an alkylbicyclobutonium ion, very similar to the cyclopropylcarbinyl cation studied by Roberts and Mazur (25). The unusual alkylbicyclobutonium ion is considered as a resonant hybrid of pyramidal structure that interconverts at different rates. The

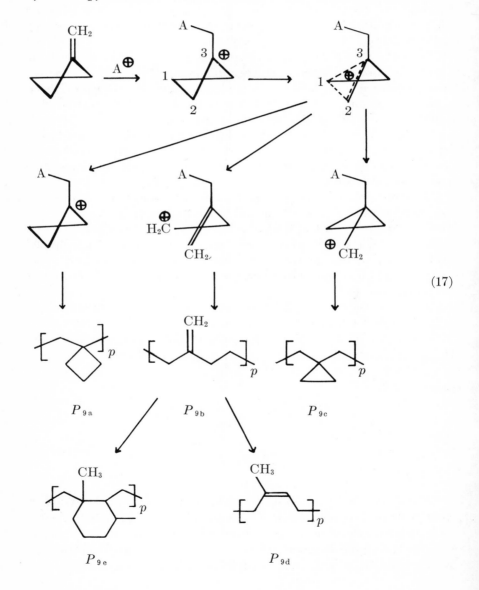

(17)

positive charge derives from relocalization of the π orbital, and the unstable cyclobutylium ion rearranges to an unusual ion by delocalization—for example, of the 3-4 covalence doublet between carbons 1-3-4. The substituent may stabilize a resonant form, thus favoring products derived from this form. Generally speaking, during polymerization, relocalization of the electrons of the alkylbicyclobutonium ion can occur by three different processes. In relocalizing, the charges give, as possible structures, P_{9a}, P_{9b}, and P_{9c}. In the presence of acid catalysts, structure P_{9b} may isomerize to structure P_{9d}, where the double bonds become trisubstituted.

Finally, it has been proved that structure P_{9d} could cyclize quite easily to structure P_{9e} because of the α hydrogens from the double bond.

With aluminum-chloride catalysts, the polymer structure is 80% cyclobutane units and about 5% cyclopropane units, the remainder being polyene units. With stannic chloride, we do not observe any cyclopropane units; the polymer consists mainly of cyclobutane units (75%). The presence of methyl groups on cyclohexane shows that the other units possess partially isomerized and cyclized polyene structures. When using titanium chloride or ethyl aluminum chloride with traces of water as cocatalyst, the polymer consists mainly of cyclobutane units—that is, 60%, the demainder being cyclized; *see* Equation 17.

POLYMERIZATION OF METHYLENECYCLOBUTANE BY TRANSITION-METAL COMPLEXES. Various transition metal halides were used: $TiCl_4$, $SnCl_4$, WCl_6, VCl_3, etc., complexed with organometallic compounds such as Et_3Al, Et_2AlCl, (*iso*-Bu)$_3$Al, and Bu_3SnH.

In the presence of $Et_3Al/TiCl_4$, polymerization of methylene cyclobutane gives a polymer having exomethylene units (P_{9b}) representing about 40% of the structure, 60% being partially cyclized.

With $Et_2AlCl/TiCl_4$, the polymer structures are different, depending on the reaction temperature. At low temperatures, the polymer is a mixture of structures P_{9a}, P_{9b}, P_{9c}, and P_{9d}. Infrared spectrography reveals the presence of P_{9a} structures by vibrations at 920 cm^{-1} and 1240 cm^{-1}. The homoallylic structure P_{9b} is demonstrated by an absorption band at 890 cm^{-1}. This polyene structure is not alone, since characteristic absorptions of trisubstituted double bonds P_{9d}, $R_1R_2C=CHR_3$, are visible with shoulders at 930 and 835-840 cm^{-1} for the trans and cis forms, respectively.

NMR spectra show a massive peak at $\delta = 4.8$ ppm, characteristic of structure P_{9d}. At $\delta = 5.2$ ppm, a slight signal indicates structure P_{9d}. The P_{9b} unit can become partially cyclized to give structure P_{9e}. When the polymer is prepared at low temperature, its bromine index is about 54%, which corresponds to a monocyclized structure P_{9e}. As the polymerization temperature increases, the intensity of the characteristic sig-

nal of methyl groups on saturated carbon atoms goes up, and that of
α protons from the double bonds diminishes in the same proportion.
The peak at $\delta = 1.5$ ppm becomes the most important (cyclohexane protons). The saturated-carbon methyl peak at $\delta = 0.9$ ppm, virtually nonexistent for low-temperature polymers, increases in size. As polymerization temperature increases, the cyclization degree increases, while the
proportion of carbon-carbon double bonds decreases (Equation 17).

Various acetylacetonates have been used with diethylaluminum
chloride as catalyst for the polymerization of methylene cyclobutane. In
this case, the NMR spectra of the polymers are very simple. Between
$\delta = 4.6$ and 4.8 ppm lies a singlet corresponding to the resonance of
vinylidene protons. Between $\delta = 1.85$ and 2.10 ppm, a signal reveals the
protons of α methylene groups from a double bond, and at $\delta = 1.25$ ppm,
a peak corresponds to protons of methylenic groups. The IR spectrum
shows no absorption at 920 cm^{-1} and 1240 cm^{-1}, which excludes the
presence of a cyclobutane structure. The structure $CH_2 = CR_1R_2$ is
indicated by vibration of the C–H bonds at 890 and 1640 cm^{-1} for
C=C stretching vibrations. A strong absorption at 1450 cm^{-1} corresponds to stretching vibrations of the C–H bonds of the methylene
groups. The absorption of methyl groups at 1370 cm^{-1} is very weak.
At 790 cm^{-1}, the CH_2 rocking vibration becomes apparent, thus confirming the succession of three –CH_2– groups. No characteristic signal
of cyclopropane structure appears. So it may be concluded that polymerization by this type of complex leads to the formation of a pure
"isopolyisoprene" form in which the carbon-carbon double bond is in
the exo position with respect to the chain.

Under a vanadium triacetylacetonate organoaluminum catalyst
$(CH_3COCH=CO-CH_3)_3V–Et_2AlCl$, methylenecyclobutane gives a polymer in which the isopolyisoprene structure (P_{9b}) is predominant—that is,
85 to 95%. However, NMR spectra show characteristic signals of cyclopropane structures between 0 and 0.2 ppm. Integration give a proportion
of 5 to 15% of this type of structure. This is confirmed by infrared
bands at 3080 and 1015 cm^{-1}.

Ternary catalysts, such as $Et_2AlCl/TiCl_4/Et_3N$ or $Et_3Al/TiCl_4/Ph_3P$
have been used. According to the various proportions of each compo-
nent of the ternary catalyst, polymers of P_{9b}, P_{9d}, and P_{9e} structures are
obtained. However, in some cases, especially when the ratio $Ph_3P/TiCl_4 = 1$, methylenecyclobutane gives a pure and linear "isopolyisoprene"
structure (P_{9b}), in which cyclized type structures and cyclopropanic
groups are excluded.

POLYMERIZATION OF BICYCLOPROPYLE. Bicyclopropyle has already
been synthesized in small yields from butadiene with a methylene iodide
and a zinc–copper couple (26, 27), and by reduction of 2,2,2',2'-tetra-

halobicyclopropyle (*28, 29*). Bicyclopropyle also has been obtained *via* a new route from butadiene by successive carbenations and reductions. The dichlorocarbene added to butadiene or vinylcyclopropane is obtained by reaction of a strong base (sodium teramylate) with chloroform, and reducing the *gem*-dichlorocyclopropyl by a sodium-hydrated methanol system. Equation 18 depicts the synthesis. The overall yield of the synthesis from butadiene is about 20%.

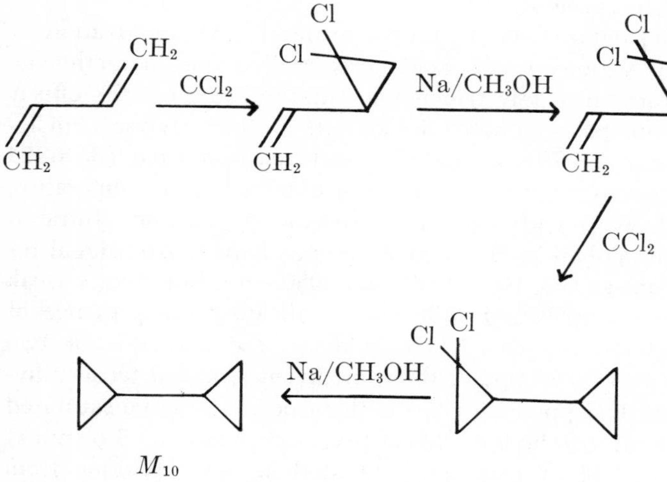

$$(18)$$

Electron diffraction studies have shown (*30*) that, in its vapor phase, the bicyclopropyle molecule possesses two isomeric conformations: a nonrigid *s*-trans form (A) and a nonrigid left form (B) (Equation 19) able to oscillate respectively from $\pm 80°$ to $\pm 18°$ around an equilibrium position.

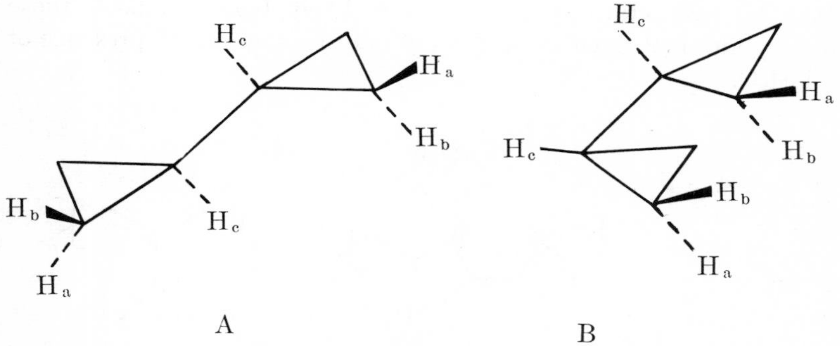

$$(19)$$

We have confirmed this point of view by testing bicyclopropyle from in its liquid phase by NMR. The existence of the two conformers is confirmed by an important splitting of the characteristic signals of the four hydrogens H_a ($\delta = -0.16$ to 0.13 ppm); the four hydrogens H_b ($\delta = 0.17$ to 0.49 ppm); and the four hydrogens H_c ($\delta = 0.57$ to 0.95 ppm).

With suitable initiators, the two conformers may yield polymers of different structures. The existence of a pseudoconjugation between the two cyclopropanes, which may be compared with those of the butadiene orbitals, theoretically favors a polymerization by simultaneous opening of the two conjugated orbitals.

CATIONIC POLYMERIZATION OF BICYCLOPROPYLE. Polymerization of bicyclopropyle by various Lewis acids was carried out in methylene chloride with a constant catalyst/monomer ratio to bring out the effects of the other parameters—namely, the nature of the catalyst ($SnCl_4$– $AlCl_3$), temperature ($-78°$ to $120°C$), and reaction time (3 to 24 hours). The highest conversion (85%) was obtained at a temperature between 70° and 80°C, with aluminum chloride as initiator. Infrared spectroscopy of the polymers obtained shows no characteristic signal for cyclopropane groups at 860, 1020, 1040, and 3080 cm^{-1}, but gives a weak broad band between 1620 and 1700 cm^{-1}, indicating the presence of tetra-substituted C=C bonds. NMR indicates the existence of two methyls (signals at $\delta = 0.90$ ppm), three methylenes, and a tertiary hydrogen ($\delta = 1.22$ to 1.27 ppm) situated in the α position from a saturated carbon. It also indicates the presence of two methylenes ($\delta = 1.80$ ppm) and a methyl ($\delta = 1.54$ to 1.60 ppm) situated in the α position from tetrasubstituted double bonds.

These results show that polymerization occurs by opening of the two cyclopropyl groups, and the existence of a C=C bond for two monomer units suggests a cyclization reaction. These lead to a polymer structure characterized by substituted cyclohexene groups in the chain. Two isomeric structures agree with these assignments (P_{10a} and P_{10b} of Equation 20). Structure (P_{10a}) is the most likely, being similar to those obtained by intramolecular cyclization of poly-1,4-isoprene in presence of Et_2AlCl–H_2O.

$$P_{10a} \qquad P_{10b} \qquad P_{10c} \tag{20}$$

Osmometric determination of molecular weights was not possible because of the low solubility of the polymer in solvents at room temperature. However, NMR spectra could be obtained by swelling the polymer in carbon tetrachloride.

POLYMERIZATION OF BICYCLOPROPYLE BY TRANSITION METAL COMPLEXES. Heterogeneous-phase polymerization of bicyclopropyle with Ziegler-Natta catalysts was carried out in n-hexane between $-30°$ and $120°C$ over 24 hours. These systems, in decreasing order of reactivity give polymers having spectral characteristic similar to those of polymers obtained by cationic initiation: $SnCl_4/EtAlCl_2$, $SnCl_4/Et_2AlCl$, $SnCl_4/Et_3Al$, $TiCl_4/Et_2AlCl$, WCl_6/Et_3Al. The same structures of cyclohexene monomer units separated by three methylenes in chain are proposed (structures P_{10a} and P_{10b} of Equation 20).

By contrast, we could give evidence for different structures when the $TiCl_4$–Et_3Al complex is used ($Ti/Al=1$). Infrared spectra show, in addition to the peaks already observed in previous polymerizations, absorption bands at 860, 1020, 1040, 1780, and 3080 cm^{-1}, characteristic of cyclopropanic groups, and bands at 770 and 780 cm^{-1}, which can be attributed to bimethylenic blocks $-CH_2-CH_2-$. NMR confirms these results. The cyclopropane protons appear at $\delta = 0.02$ and 0.44 ppm, and methylenic hydrogens of a linear chain are characterized by a signal at $\delta = 1.27$ ppm. These spectrographic data and the similarities with those of polyvinylcyclopropane (*31*) suggest that the chain structure is composed of units of undescribed linear structure consisting of methylene groups in the main chain of cyclopropane groups in side chain (structure P_{10c}). A study of this type of polymerization shows that over a wide range of reaction temperatures ($20°$ to $120°C$), the linear form predominates (from 55 to 70%), 40 to 30% being cyclized (P_{10a}, P_{10b}, P_{10c}); *see* Equation 20.

Cationic initiation thus opens the two cyclopropane structures to give polymers in which the trimethylcyclohexene rings are separated by three methylenes in the chain. With $TiCl_4$–Et_3Al, the polymerization process is different, and the reaction involves the opening of one ring of the monomer. The polymer obtained has a molecular weight of about 5000, and consists of blocks identical to the foregoing structures (P_{10a} and P_{10b}) and of blocks formed of monomer units with pendant cyclopropane groups (P_{10c}).

The results obtained with bicyclopropyle are quite different from those obtained with spiropentane, which, whatever the type of catalyst used, yields cyclized blocks similar to those obtained with poly-1,4- or 3,4-isoprene.

Polymerization of gem-*Dihalocyclopropanic Systems*

The *gem*-dihalocyclopropane type of structure, considered highly stable, is attacked by certain electrophilic reagents or simply by heating, and rearranges by ring opening. Thus, with Lewis-acid catalysts, the dihalocyclopropane derivative opens with formation of a haloallylic carbo cation (*32*) which, in presence of a nucleophilic ion, gives either a haloallylic derivative or a 1,3-diene.

Pyrolysis of dihalocyclopropanes was studied along with the effects of electrophilic reagents, and confirms the foregoing results (*33-37*). In many cases, those authors observed that polymeric residues, in addition to allylic and diene-type products, were present at the end of pyrolysis. The formation of these polymers confirms the hypothesis that the dihalocyclopropanes are monomers that can be polymerized either by cationic processes or by the action of transition-metal complex catalysts.

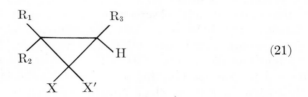

$$(21)$$

1,1-dimethyl-2,2-dichlorocyclopropane (M_{11})

($R_1 = R_2 = CH_3$; $R_3 = H$; $X = X' = Cl$)

1,1,3-trimethyl-2,2-dichlorocyclopropane (M_{12})

($R_1 = R_2 = R_3 = CH_3$; $X = X' = Cl$)

1,1-dimethyl-2-chloro-2-bromocyclopropane (M_{13})

($R_1 = R_2 = CH_3$; $R_3 = H$; $X = Cl$; $X' = Br$)

Polymerization of Alkyl gem-**Dihalocyclopropanes.** These monomers (Equation 21) are prepared by adding dihalocarbenes to the corresponding ethylenic compounds—that is, isobutene for M_{11} and M_{13}, and 2-methyl-2-butene for M_{12}. The carbenes are produced by basic α elimination of haloforms in the presence of a strong base such as sodium teramylate (*15*), or by basic α elimination of trichloroacetic esters. Infrared data of these compounds show absorption bands at 3040 cm and 1020 cm^{-1} for cyclopropyl groups. NMR gives signals at 1.2 and 1.3 ppm.

CATIONIC POLYMERIZATION. Monomers M_{11}, M_{12}, and M_{13} were treated with Lewis-acid catalysts such as $AlCl_3$, $TiCl_4$, and $SnCl_4$. The polymers obtained give absorption bands (1650 and 850 cm^{-1}) and NMR peaks (1.1 and 5.5 ppm), indicating the presence of a chlorinated double bond and methyl group protons on a saturated carbon atom, respectively. The infrared bands corresponding to cyclopropane have completely disappeared. Microanalytical determinations give an empirical formula

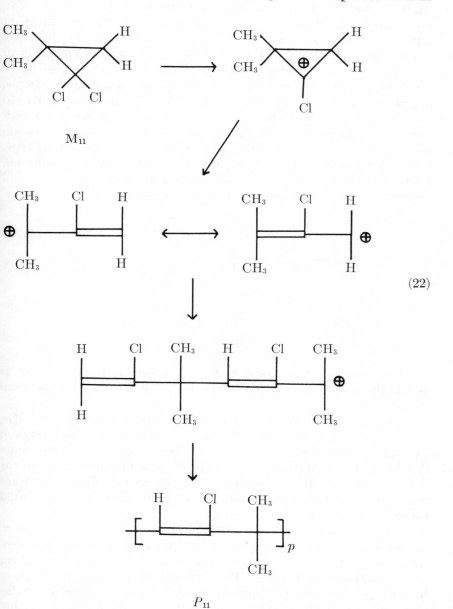

(22)

that indicates elimination of one hydrohalide molecule per monome unit of the chain.   These results suggest the opening of a three-carbo ring and a dehydrohalogenation's occurring during the polymerization.

The following mechanism may be proposed (in the case of M_{11}) attack by the cationic initiator involves the tricentric ring opening.   Th carbo cation formed may be drawn in several resonant forms.   Attack o a further monomer molecule gives a polymer of structure P_{11}, having chlorinated double bond and a *gem*-dialkyl group.   This structure is i agreement with the experimental results (Equation 22).

However, the reaction may develop in other ways.   For example deprotonation of a methyl group in the chloroallylium group can yiel 2-methyl-3-chloro-1,3-butadiene, which, under the experimental condi tions used, can easily polymerize to give several structures (Equatio 23), depending on whether polymerization is of the 1,2,3,4, or 1,4 type NMR spectrography makes it possible to eliminate this eventuality be cause of the absence of peaks corresponding to protons of an α methy group from the double bond and methylene group in the chain.

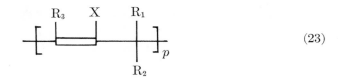

$$\tag{23}$$

$$P_{11} \ (R_1 = R_2 = CH_3; \ R_3 = H; \ X = X' = Cl)$$

$$P_{13} \ (R_1 = R_2 = CH_3; \ R_3 = H; \ X = Cl)$$

The reaction temperature is quite important.   At 20°C, there i practically no reaction; at 80°C, the degree of conversion is betwee 15 and 20%.

POLYMERIZATION BY TRANSITION-METAL COMPLEX CATALYSTS. M_{11} M_{12}, and M_{13} have been polymerized by $Et_3Al/TiCl_4$ catalysts betwee 50° and 80°C in *n*-hexane, the reaction times ranging from a few hour to several days.   The polymers obtained have the same structure a those obtained by cationic polymerization.   By analogy with mechanism proposed in the literature *(38, 39)*, the structure shown in Equation 2 may be proposed for the active center.

The formation of HX in a stoichiometric amount with respect t the monomer probably involves the disappearance of many active cen ters.   This explains the relatively large amount of catalyst required fo polymerization to take place.   Furthermore, the polymerization degre remains low, and the polymers obtained have low molecular weight (for example, 1000 to 3000).   Temperature is still a decisive facto

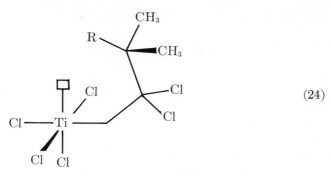

(24)

Conversion degrees become appreciable above 50°C. In addition, the conversion degree increases as the amount of catalyst increases.

In summary, the polymerizability of *gem*-dihalocyclopropanes decreases from M_{11} to M_{12}, and becomes zero in the case of 1,1,2,2-tetramethyl-3,3-dichlorocyclopropane, which does not react in the presence of the polymerization catalysts used. The type of substitution of cyclopropane therefore seems to be an important factor. In all cases, polymerization implies opening of the three-carbon ring and dehydrohalogenation of each unit. Polymerization of these compounds is favored by high temperatures, above 20°C, and catalyst concentrations much higher than in olefin polymerizations. The molecular weights are of the order of 1000 to 3000, and the oligomers, soluble in the usual organic solvents, are white powders melting above 150°C.

Polymerization of Dihalocyclopropane with an Adjacent Phenyl Group. The monomers shown in Equation 25 are preparated by adding

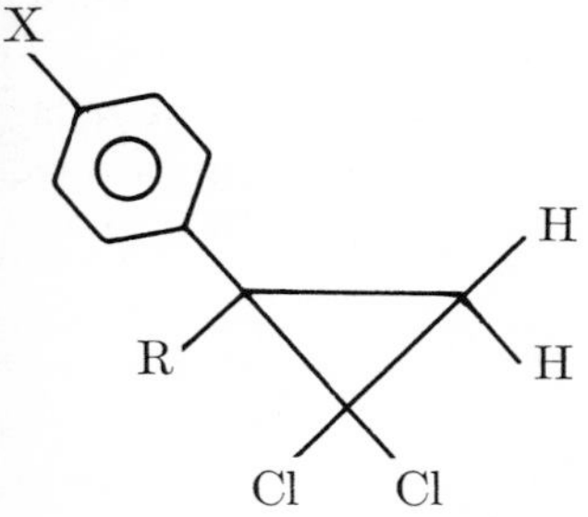

M_{14} (R = H; X = H)

M_{15} (R = CH$_3$; X = H)   (25)

M_{16} (R = CH$_3$; X = Cl)

2-phenyl-1,1-dichlorocyclopropane (M_{14})
(R = H; X = H)

1-phenyl-1-methyl-2,2-dichlorocyclopropane (M_{15})
(R = CH$_3$; X = H)

1-p-chlorophenyl-1-methyl-2,2-dichlorocyclopropane (M_{16})
(R = CH$_3$; X = Cl)

dichlorocarbenes to styrene, α-methylstyrene, and p-chloro-α-methylstyrene.

CATIONIC POLYMERIZATION. In the compounds of Equation 25, the pseudoconjugation between the phenyl and the cyclopropane is maximum when the two rings are perpendicular (*40*). Polymerization, cationic or by transition-metal complex catalysts, indicates the participation of the phenyl group.

Cationic polymerization yields oligomers that have a structure similar to that of the polymers previously described. Spectrographic studies and microanalytical results indicate the disappearance of the three-carbon ring and elimination of one molecule of HCl per molecule of monomer. These results agree with a structure that would result from a polymerization mechanism similar to that proposed in the previous case. However, the presence of an NMR peak (1.6 ppm) indicates participation of the phenyl group during polymerization in para position, giving the structure described in Equation 25. With the same reaction conditions as with monomers M_6 and M_7, monomer M_8 (in which the para position is substituted by a chlorine atom) does not give any polymer.

POLYMERIZATION BY ZIEGLER-NATTA CATALYSTS. Under heterogeneous Ziegler-Natta type catalysis with Et_3Al–$TiCl_4$, Et_3Al–$SnCl_4$, Et_2AlCl–$TiCl_4$, Et_2AlCl–$SnCl_4$, $EtAlCl_2$–$SnCl_4$, $EtAlCl_2$–$TiCl_4$, i-Bu_3Al–$TiCl_4$, and i-Bu_3Al–$SnCl_4$ in hexane, the polymers have the same characteristics as those obtained with cationic catalysts. The effects of various polymerization parameters were studied, (concentration of catalyst, Al/M ratio, temperature, polymerization time, etc.) Temperature is a factor that favors an increase in the conversion degree, with a maximum at 80°C. The polymers obtained from M_{14} are only slightly soluble in common organic solvents, whereas polymers obtained from M_{15} are highly soluble in the same solvents. This difference may be attributed to interchain linkages that cannot happen in the case of M_7 because of the presence of methyl groups. The molecular weights are low; $\overline{M}_n \simeq 2000$ (Equation 26).

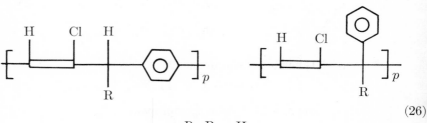

$$(26)$$

$$P_{14} \; R = H$$

$$P_{15} \; R = CH_3$$

Polymerization of 2-Methyl-2-vinyl-1,1-dichlorocyclopropane. The case of vinylcyclopropane compounds is of particular interest because of the conjugation between the cyclopropane and the C=C double bond, and the analogy between these compounds and 1,3-dienes (*41, 42*). The polymerization of 2-methyl-2-vinyl-1,1-dichlorocyclopropane (M_{17}) was therefore studied. This monomer is prepared by adding dichlorocarbene to isoprene.

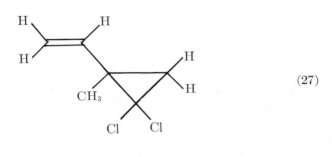

$$(27)$$

$$M_{17}$$

CATIONIC POLYMERIZATION. Use of $TiCl_4$, $SnCl_4$, and WCl_6 as catalysts yields oligomers with average molecular weights of about 5000. The conversion degree is a function of temperature, the maximum being at 80°C. Analysis indicates an empirical formula corresponding to the removal of one molecule of HCl per monomer unit, as in the previous cases. Infrared spectroscopy and NMR confirm the disappearance of the cyclopropane group. A 1,5-type of polymerization (*43-46*) can be eliminated. On the other hand, a 1,2-polymerization of the double bond alone cannot be assumed since the *gem*-dichlorocyclopropyl group has completely disappeared. This is proved by spectroscopic data. The polymers obtained give infrared absorption bands at 1615 cm⁻¹ and 890 cm⁻¹, and NMR peaks at 0.95 and 1.2 ppm.

In fact, a cyclopropylcarbinyl ion is formed as the result of the attack by the cationic initiator. This carbo cation may rearrange, through intermediate bicyclobutonium ions, to give cyclobutylium or allylcarbinyl ion. The basic units obtained from such rearrangements have a structure that is either cyclobutenic or allylic. The predominant structures of the polymers are P_{17a} and P_{17b} of Equations 28 and 29.

POLYMERIZATION BY ZIEGLER-NATTA CATALYSTS. Polymerization by transition-metal complex catalysts gives oligomers of the same structures as those obtained cationically, with distinctly higher conversion degrees (Equation 29).

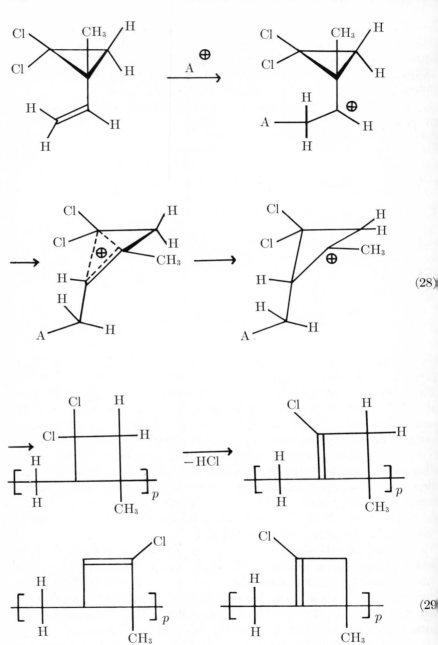

$$(28)$$

$$(29)$$

P_{17a} P_{17b}

   The reactivity of this vinylcyclopropane compound toward cationic or Ziegler-Natta initiators is greater than that of purely cyclopropane-type monomers.   This stems from the conjugation of the two unsaturated systems; the results obtained agree with work carried out on the polymerization of vinyl cyclopropane itself.

   Polymerization of *gem*-Dihalobicyclo[*n*.1.0]alkanes.   These monomers are prepared by addition of dichlorocarbene to the corresponding cyclenes (M_{18} to M_{25} of Equation 30).   The tricentric ring-opening reactions of the *gem*-dihalobicyclic compounds has been studied both by using electrophilic reactants and by pyrolysis.   In all cases, opening of the three-carbon ring is made by rupture of the bond common to both rings.

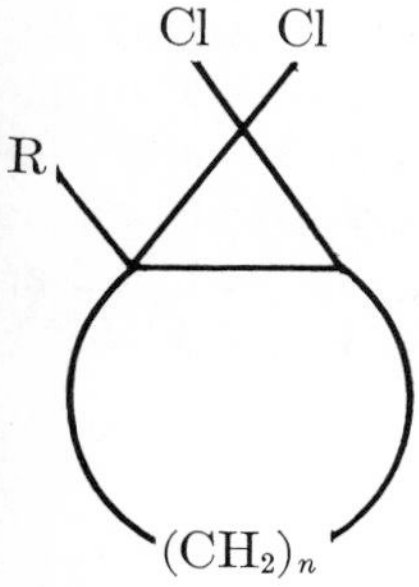

$$
\begin{array}{llll}
& n = 4 & M_{18} & (R = H) \\
& & M_{19} & (R = CH_3) \\
& n = 5 & M_{20} & (R = H) \\
& & M_{21} & (R = CH_3) \\
& n = 6 & M_{22} & (R = H) \\
& & M_{23} & (R = CH_3) \\
& n = 7 & M_{24} & (R = H) \\
& & M_{25} & (R = CH_3)
\end{array}
\tag{30}
$$

   CATIONIC POLYMERIZATIONS.   Catalysts such as $TiCl_4$, $SnCl_4$, and $AlCl_3$ give polymers whose spectroscopic data show NMR signals at 0.9, 1.6, and 2.1 ppm, corresponding respectively to methyl protons on saturated carbon atom, to methylene protons on the ring, and to one α-methylene group from C=C double bond.   The integration ratio gives

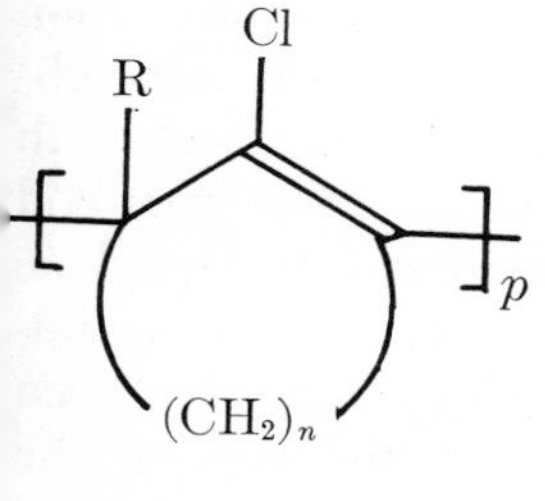

$$
\begin{array}{llll}
n = 3 & P_{18} & (R = H) \\
n = 4 & P_{19} & (R = CH_3) \\
& P_{20} & (R = H) \\
n = 5 & P_{21} & (R = CH_3) \\
& P_{22} & (R = H) \\
n = 6 & P_{23} & (R = CH_3) \\
n = 9 & P_{24} & (R = H) \\
n = 10 & P_{25} & (R = CH_3)
\end{array}
\tag{31}
$$

a proportion of one methyl group per monomer unit of polymer. Microanalytical results give evidence of the loss of one HCl molecule per monomer unit. According to these results, the structure of this polymer is as shown in Equation 31.

The importance of the size of the ring fused to the cyclopropane is demonstrated by systematic studies carried out with a great number of initiators. In particular, although the size of the ring does not appear to affect the conversion degree, it seems that the molecular weight of the polymer obtained increases as ring size decreases.

POLYMERIZATION BY ZIEGLER-NATTA CATALYSTS. The Ziegler-Natta catalysts have been used with the same conditions as those described above. Polymers obtained under such conditions present the same structures as those of polymers obtained by cationic polymerization.

Polymerization of Bicyclo[6.1.0]non-4-ene and 9,9-Dihalobicyclo-[6.1.0]non-4-ene. Bicyclo[6.1.0]non-4-ene and 9,9-dihalobicyclo[6.1.0]-non-4-ene (M_{26}, Equation 32) have the special feature of having two unsaturated sites that can react separately or simultaneously—that is, one C=C double bond and a cyclopropane (47-50); 9,9-dihalobicyclo[6.1.0]-non-4-ene was synthesized by adding dihalocarbene to 1,5-cyclooctadiene (yield 56%). Reduction of the dihalocyclopropane group with a Na/hydrated methanol system yielded bicyclo[6.1.0]non-4-ene (yield 85%).

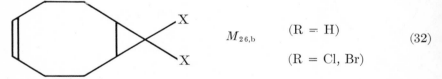

$$M_{26,b} \qquad (R = H) \qquad\qquad (32)$$
$$(R = Cl, Br)$$

In the presence of cationic initiators, these monomers polymerize *via* a transannular mechanism to give polymers of structures P_{26a1} and P_{26b1} for the bicyclo[6.1.0]non-4-ene, and dihalo-9,9-bicyclo[6.1.0]non-4-ene, respectively (Equation 33).

In the presence of Ziegler-Natta catalysts, the two sites participate in the reaction, but the structures of the polymers obtained are slightly different: P_{26a2} and P_{26b2}, respectively.

However, in presence of WCl_6–$EtAlCl_2$ in benzene at room temperature (molar ratio M/W=800, molar ratio Al/W=8), 50% conversion polymers having molecular weights between 2500 and 6500 were obtained. The NMR spectra of these polymers show some similarity with those of the monomer. In both cases, two signals at $\delta = 5.35$ ppm and 2.1 ppm indicate that a double bond is situated between two methylene groups: at $\delta = 0.62$ ppm and -0.26 ppm, two peaks appear corresponding to cyclopropane protons. The signal at $\delta = 1.35$ ppm can be given by methylene α protons from the cyclopropane. In the case of the

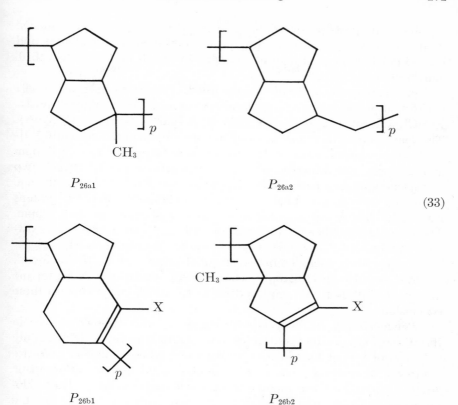

P_{26a1} P_{26a2}

(33)

P_{26b1} P_{26b2}

monomer, because of the effect of the ring, this peak is divided in two at $\delta = 2.1$ ppm and 1.30 ppm. These results and the integration of the polymer spectra suggest a 1,4-polybutadiene type structure P_{26a3}, in which one out of two double bonds is replaced by a cyclopropane group (Equation 34).

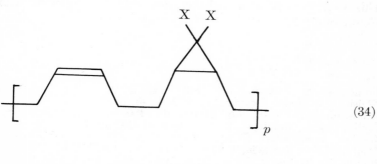

(34)

X = H P_{26a3}

X = Cl, Br P_{26b3}

Infrared spectroscopy confirms this hypothesis. The cyclopropane groups are clearly characterized by peaks at 3070 cm and 1030 cm^{-1}, and the 1,4-polybutadiene structure is verified by bands at 1660, 1410, 1310, and 1080 cm^{-1}.

Polymerization of 9,9-dichlorobicyclo[6.1.0]non-4-ene gives a 30% conversion polymer having a molecular weight of 2000 (catalyst: WCl$_6$– EtAlCl$_2$; solvent: benzene; molar ratio M/W = 300, molar ratio Al/W = 8). The spectroscopic data may be compared with those of the bicyclo[6.1.0]-non-4-ene polymers. The NMR spectra show two peaks at $\delta = 5.45$ ppm and 2.2 ppm, indicating the existence of a double bond between two methylene groups; a peak at $\delta = 1.5$ ppm may be attributed to methylene α protons from a *gem*-dihalocyclopropane (51). The tertiary protons carried by this group are characterized by a shoulder at $\delta = 1.3$ ppm. These results, together with integration of the spectra, suggest the existence of structure P_{26b3} similar to structure P_{26a3} in Equation 34.

Furthermore, microdeterminations of chlorine (Calcd., 37.12%; found, 35.26%) and IR spectroscopy (characteristic bands for cyclopropane groups at 3070 and 1030 cm^{-1} and for C–Cl at 805 cm^{-1}) confirm these conclusions.

Polymerization of bicyclo[6.1.0]non-4-ene and 9,9-dichlorobicyclo [6.1.0]non-4-ene in the presence of WCl$_6$–EtAlCl$_2$ catalyst yields mainly polymers of 1,4-polybutadiene-type structures, whose monomer units are rigorously alternating, containing a double bond and a cyclopropane group separated by two methylene groups linked together. The cyclopropanes, with or without substituents, take virtually no part in the polymerization process. More information has already been published (52-58).

Acknowledgment

This work was done in collaboration with F. Clouet, G. Clouet, J. C. Soutif, and J. P. Villette.

Literature Cited

1. Bernett, W. A., *J. Chem. Educ.* (1967) **44**, 17.
2. Walsh, A. D., *Trans. Faraday Soc.* (1949) **45**, 179.
3. Hoffmann, R., *J. Chem. Phys.* (1964) **40**, 2480.
4. Coulson, C. A., Moffitt, W. E., *Phil. Mag.* (1949) **49**, 1.
5. Ingrham, L. L., "Steric Effects in Organic Chemistry," (1965).
6. Coulson, C. A., Goodwin, T. H., *J. Chem. Soc.* (1962) 1285.
7. Randic, M., Maksic, Z., *Theor. Chim. Acta* (1965) 3, 59.
8. Tipper, C. F. H., Walker, D. A., *J. Chem. Soc.* (1959) 1352.
9. Brossas, J., Thèse de Doctorat d'Etat, Laboratoire de Chimie Macromoléculaire, Collège Scientifique Universitaire du Mans (1969).

10. Ketley, A. D., *J. Polym. Sci.* (1963) **B1**, 313.
11. Naegele, W., Haubenstock, H., *Tetrahedron Lett.* (1965) **48**, 4283.
12. Aoki, S., Harita, Y., Otsu, J., Imoto, M., *Bull. Soc. Chim. Jap.* (1966) **39**, 889.
13. Yanagita, M., Yamado, A., Suzuki, M., *Rika Gaku Kenkyusho Hokoku* (1967) **37**, 429.
14. Simmons, H. E., Smith, R. D., *J. Amer. Chem. Soc.* (1959) **81**, 4256.
15. Hine, J., Ehrenson, S. J., *J. Amer. Chem. Soc.* (1958) **80**, 824.
16. Doering, W. E., Hoffmann, A. K., *J. Amer. Chem. Soc.* (1954) **76**, 6162.
17. Natta G., Danusso, F., "Stereoregular Polymers and Stereospecific Polymerizations," Vol. 1. p. 296, 1967.
18. Ketley, A. D., Ehrig, R. J., *J. Polym. Sci.* (1964) **A2**, 4461.
19. Pinazzi, C., Brossas, J., *Die Makro. Chem.* (1969) **122**, 105.
20. Gaylord, N. G., Kossler, I., Stolka, M., Vodehnal, J., *J. Polym. Sci.* (1964) **A2**, 3969.
21. Stolka, M., Vodehnal, J., Kossler, I., *J. Polym. Sci.* (1964) **A2**, 987.
22. Golub, M. A., Heller, J., *Can. J. Chem.* (1963) **41**, 937.
23. Fanta, G. F., Ph.D. thesis, University of Illinois (1960).
24. Pinazzi, C., Reyx, D., *Bul. Soc. Chim. Fr.* (1972) 3930.
25. Roberts, J. D., Mazur, R. H., *J. Amer. Chem. Soc.* (1951) **73**, 2509.
26. Overberger, C. G., Malek, G. W., *J. Org. Chem.* (1963) **28**, 867.
27. Wittig, G., Wingler, F., *Ann. Chem.* (1964) **97**, 2139.
28. Schrumpf, G., Luttke, W., *Liebigs Ann. Chem.* (1969) **730**, 100.
29. Skattebøl, L., *J. Org. Chem.* (1964) **29** (**10**), 2951.
30. Bastiansen, O., de Meijere, A., *Angew. Chem. Internat. Edit.* (1966) **5** (**1**), 124.
31. Ermakova, I. I., Kropacheva, E. M., Kol'tsov, A. I., Dolgoplosk, B. A., *Vysokomol. Soedin.* (1969) **11** (**7**), 1639.
32. Skattebøl, L. Boulette, B. *J. Org. Chem.* (1966) 31, 81.
33. Neureiter, N. P., *J. Org. Chem.* (1959) **24**, 2044.
34. Ellis, R. J., Frey, H. M., *J. Chem. Soc.* (1964) 959.
35. Winberg, H. E., *J. Org. Chem.* (1959) **24**, 264.
36. Robinson, G. C., *J. Org. Chem.* (1964) **29**, 3433.
37. Ketley, A. D., Berlin, A. J., Gorman, E., Fischer, L. P., *J. Org. Chem.* (1966) **31**, 305.
38. Natta, G., Mazzanti, G., *Tetrahedron* (1960) **8**, 86.
39. Cossee, P., *J. Catal.* (1964) **3**, 80.
40. Roberston, W. W., Music, J. F., Matsen, F. A., *J. Amer. Chem. Soc.* (1950) **72**, 5260.
41. Overberger, C. G., Borchert, A. E., *J. Amer. Chem. Soc.* (1969) **82**, 1007.
42. Flowers, M. C., Frey, H. M., *J. Chem. Soc.* (1961) 3547.
43. Takahashi, T., Yamashita, I., Miyakawa, T., *Bull. Chem. Soc. Jap.* (1964) **37**, 131.
44. Takahashi, T., Yamashita, I., *J. Polym. Sci.* (1965) **B-3**, 251.
45. Takahashi, T., Yamashita, I., *Kogyo Kagaku Zasshi* (1965) **68**, 869.
46. Takahashi, T., *J. Polym. Sci., Part A-1*, (1968) **6**, 403.
47. Natta, G., Dall'asta, G., Bassi, I. W., Carella, G., *Makromol. Chem.* (1966) **91**, 87.
48. Calderon, N., Ofstead, E. A., Judy, W. A., *J. Polym. Sci., Part A-1* (1967) **5**, 2209.
49. Scott, K. W., Calderon, N., Ofstead, E. A., Judy, W. A., Ward, J. P., ADVAN. CHEM. SER. (1969) **91**, 26, 399.
50. Ofstead, E. A., *Synth. Rub. Symp., London* (1969).
51. Pinazzi, C. P., Reyx, D., unpublished paper.
52. Pinazzi, C. P., Levesque, G., Reyx, D., Brosse, J. C., Pleurdeau, A., ADVAN. CHEM. SER. (1969) **91**, 27, 419.

53. Pinazzi, C. P., Pleurdeau, A., Brosse, J. C., *Die Makromol. Chem.* (1971) **142**, 259.
54. Pinazzi, C. P., Brosse, J. C., Pleurdeau, A., *Die Makromol. Chem.* (1971) **142**, 273.
55. Pinazzi, C. P., Brosse, J. C., Pleurdeau, A., *Die Makromol. Chem.* (1971) **144**, 155.
56. Pinazzi, C. P., Brossas, J., *Die Makromol. Chem.* (1971) **147**, 15.
57. Pinazzi, C. P., Brossas, J., Clouet, G., *Die Makromol. Chem.* (1971) **148**, 81.
58. Pinazzi, C. P., Legeay, G., Brosse, J. C., *C. R. Acad. Sci., Paris* (1971) **273** (**C**), 797.

RECEIVED April 13, 1972.

Polymerization of Pivalolactone

N. R. MAYNE

Koninklijke/Shell-Laboratorium, Amsterdam, The Netherlands

Pivalolactone (α,α-dimethyl-β-propiolactone) can be polymerized via a "living" anionic mechanism to yield a linear polyester exhibiting properties suitable for use as a textile fiber or as an engineering plastic. Two process for the polymerization have been successfully developed by Shell to the pilot-plant stage. The first is a continuous melt-bulk process in which the monomer is rapidly polymerized in a gear-pump reactor under almost adiabatic conditions. The molten polymer is directly degassed, cooled, and cut into nibs. The second process involves polymerization in a slurry system with a heterogeneous initiator. The initiator is, in fact, "living" low-molecular-weight polymer, and yields a free-flowing powder with a high bulk density. Development and chemical background of these two processes are described here and some typical mechanical properties of the product are briefly discussed.

Pivalolactone (α,α-dimethyl-β-propiolactone) is a colorless liquid that can be prepared from pivalic acid *via* chlorination and subsequent ring closure of the sodium salt of the chloroacid.

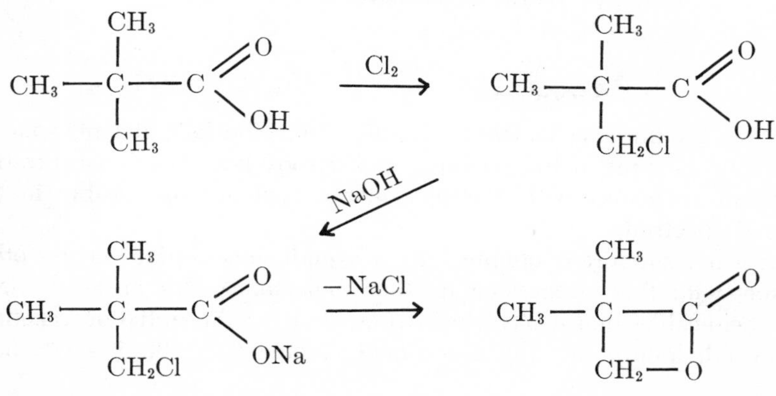

Because of the highly strained ring system, polymerization readily occurs *via* ring opening, yielding a linear polyester that exhibits properties suitable for use as a textile fiber or as an engineering plastic.

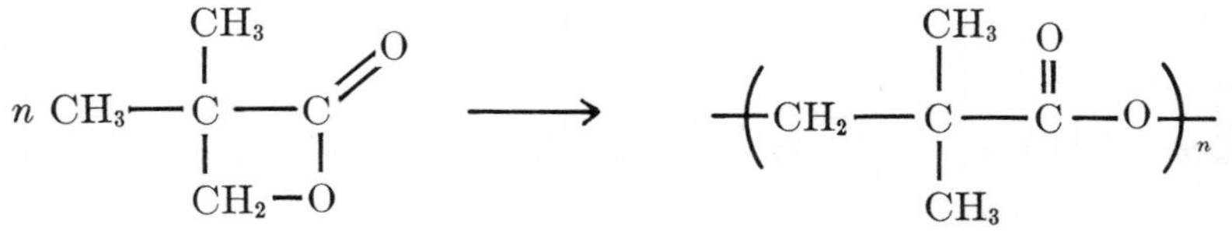

The reaction is exothermic, with a heat of polymerization of 77 kJ/mole (18.4 kcal/mole). This value includes the heat of crystallization of 12.1 kJ/mole (2.9 kcal/mole).

Before discussing in detail the two polymerization processes Shell has developed up to the pilot-plant stage, we will consider here the chemistry of the polymerization reaction.

Polymerization Reaction

The polymerization of pivalolactone can be initiated by a variety of nucleophiles. For example, tributylphosphine (TBP) attacks the molecule exclusively at the beta position, causing ring opening and propagation *via* a carboxylate ion. The high stability of this growing macrozwitterion results in a "living" polymerization system.

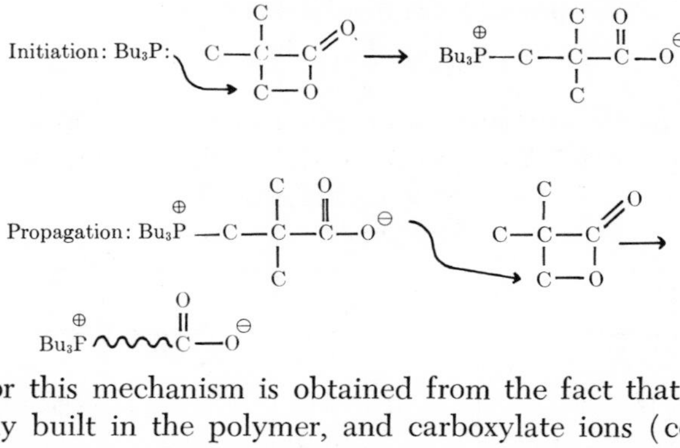

Evidence for this mechanism is obtained from the fact that phosphorus is chemically built in the polymer, and carboxylate ions (concentration of which correlates with the phosphorus content) are visible in the infrared spectrum.

Further support is obtained from experiments carried out by other workers with the parent lactone, β-propiolactone. For instance, NMR and chemical evidence have been presented for the initiation reactions with triethylphosphine (1) and trimethylamine (2). Studies (3) have also been performed on macrozwitterions prepared by initiation with the betaine $(CH_3)_3\overset{\oplus}{N}CH_2COO^{\ominus}$.

$$(CH_3)_3\overset{\oplus}{N}-CH_2-\overset{\overset{\displaystyle O}{\|}}{C}-\overset{\ominus}{O} +_n \;\; \begin{matrix} CH_2-C\diagup^{\displaystyle O} \\ | \quad\quad | \\ CH_2-O \end{matrix} \longrightarrow$$

$$(CH_3)_3\overset{\oplus}{N}-CH_2-\overset{\overset{\displaystyle O}{\|}}{C}-O\!\!\left(\!CH_2-CH_2-\overset{\overset{\displaystyle O}{\|}}{C}-O\!\right)_{\!\!n-1}\!\!\!CH_2CH_2\overset{\overset{\displaystyle O}{\|}}{C}-O^{\ominus}$$

In addition to infrared and NMR data confirming the structure of the macrozwitterion, electrophoresis has shown that quaternary ammonium cations and carboxylate anions are present in the same polymer chain.

Objections to the zwitterion polymerization mechanism because of the high Coulombic energy of charge separation are not necessarily valid because the propagating chains need not be linear, with increasing distance between the two ionic end groups. They may be cyclized or paired with another chain throughout the polymerization.

Unlike many other polymerization reactions, chain transfer with monomer cannot occur. However, many compounds (proton donors in particular) function as chain-transfer agents when added to the system. In general, the mechanism of the chain-transfer reaction can be depicted this way:

$$Bu_3\overset{\oplus}{P}\!\!\sim\!\!\sim\!\!\sim\!\!\overset{\overset{\displaystyle O}{\|}}{C}-O^{\ominus} + RX \;\rightleftharpoons\; Bu_3\overset{\oplus}{P}\!\!\sim\!\!\sim\!\!\sim\!\!\overset{\overset{\displaystyle O}{\|}}{C}-OR + X^{\ominus}$$

$$X^{\ominus} + \begin{matrix} C \\ | \\ C-C-C\diagup^{\displaystyle O} \\ | \\ C-O \end{matrix} \longrightarrow \begin{matrix} C \;\; O \\ | \quad \| \\ X-C-C-C-O^{\ominus} \\ | \\ C \end{matrix}$$

$$\begin{matrix} C \;\; O \\ | \quad \| \\ X-C-C-C-O^{\ominus} \\ | \\ C \end{matrix} + \begin{matrix} C \\ | \\ C-C-C\diagup^{\displaystyle O} \\ | \\ C-C \end{matrix} \longrightarrow X\!\!\sim\!\!\sim\!\!\overset{\overset{\displaystyle O}{\|}}{C}-O^{\ominus}$$

$$R = H, \;\; X = RCOO^{\ominus}, \;\; C_6H_5O^{\ominus}, \;\; CH_3-\overset{\overset{\displaystyle O}{\|}}{C}-\overset{\ominus}{CH}-\overset{\overset{\displaystyle O}{\|}}{C}-OC_2H_5, \text{ etc.}$$

Although the system can essentially be described as a "living" polymerization system, termination reactions are possible. These occur especially at high temperatures under the conditions encountered on processing the plastic in an extruder.

$$\sim\!\!\sim\!\!\sim\overset{\overset{\displaystyle O}{\|}}{C}\!\!-\!\!\overset{\ominus}{O}\ \ \overset{\oplus}{PBu_3}\!\!\sim\!\!\sim \longrightarrow \sim\!\!\sim\!\!\sim\overset{\overset{\displaystyle O}{\|}}{C}\!\!-\!\!OBu + PBu_2\!\!\sim\!\!\sim$$

$$\text{or} \quad \sim\!\!\sim\!\!\sim\overset{\overset{\displaystyle O}{\|}}{C}\!\!-\!\!Bu + \overset{\overset{\displaystyle O}{\|}}{P}Bu_2\!\!\sim\!\!\sim$$

These reactions have been reported for phosphonium carboxylates (4, 5). This is, in fact, one of the major factors contributing to the high thermal stability of polymers prepared with phosphines as initiators. A competing reaction is depolymerization *via* an unzipping reaction.

$$\sim\!\!\sim\!\!\sim\overset{\overset{\displaystyle O}{\|}}{C}\!\!-\!\!O\!\!-\!\!C\!\!-\!\!\overset{\overset{\displaystyle C}{|}}{\underset{\underset{\displaystyle C}{|}}{C}}\!\!-\!\!\overset{\overset{\displaystyle O}{\|}}{C}\!\!-\!\!\overset{\ominus}{O} \longrightarrow \sim\!\!\sim\!\!\sim\overset{\overset{\displaystyle O}{\|}}{C}\!\!-\!\!\overset{\ominus}{O}$$

$$+ \left[\begin{array}{c} C \\ | \\ C\!\!-\!\!C\!\!-\!\!C \\ | \quad \\ C\!\!-\!\!O \end{array} \right] \longrightarrow CO_2 + C\!\!=\!\!C\big\langle^{\displaystyle C}_{\displaystyle C}$$

The lactone monomer can be recovered from the thermal degradation reaction if removed quickly (6). As other workers have shown (7), cyclic oligomers are also evolved, but we have found that these, in turn, are decomposed into carbon dioxide and isobutene (*via* the lactone) when carboxylate ions are present. When metal salts are used as initiators, no termination reactions are possible and, consequently, the polymers have an extremely poor thermal stability.

In our search for a practical and attractive commercial process for the polymerization of pivalolactone, we had to consider some constraints (Table I).

Table I.   Process Constraints

Monomer	*Polymer*
b.p. 160°C	m.p. 240°C
dec. > 250°C	$\Delta H = 77$ kJ/mole
Low viscosity	High viscosity

In principle, the polymerization can be carried out either in bulk or with a diluent present. For a bulk process, the most attractive route would be one in which the polymer is produced in the molten state so that it could be fed directly to an extruder.

Melt-Bulk Polymerization Process

Considering the process constraints in Table I, the design of a melt-bulk polymerization process (8) must take four factors into account:

(a) To produce polymer in the molten state, the reaction temperature should be higher than 240°C. Pressure must then be applied to prevent monomer loss by evaporation.

(b) In view of appreciable monomer decomposition at these temperatures, it is essential that the polymerization reaction proceed extremely rapidly.

(c) A rapid polymerization would make heat removal from the viscous polymer mass difficult, leading to an almost adiabatic reaction.

(d) The adiabatic temperature rise should not be so high as to make the whole process impractical because of excessive monomer decomposition.

It was, therefore, important to know the adiabatic temperature rise expected for a 100% yield. Using the heat of polymerization and specific heat data, a value of about 330°C was calculated from the relation $\Delta H = C_p \cdot dT$, and using measured and extrapolated values of C_p at different temperatures for both monomer and polymer. This 330°C figure was confirmed by results obtained from adiabatic batch polymerizations

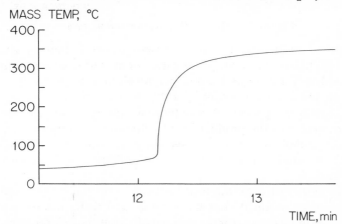

Figure 1.   Adiabatic bulk polymerization of pivalolactone

carried out in an autoclave; a maximum temperature of 350°C was re-corded (Figure 1). There is a dramatic increase in polymerization rate at about 70°C, and the maximum temperature is reached within one minute.

To obtain a continuous polymerization system, transport of the vis-cous polymer is required. Furthermore, the initiated monomer should not polymerize upstream of the reaction zone. These requirements were met by continuously injecting TBP-initiated monomer under pressure into a heated gear pump. The monomer rapidly polymerizes between the teeth of the gear wheels, and the viscous polymer leaving the gear pump is passed into an extruder, where the volatile decomposition prod-ucts are vented off. The polymer strands are then cooled and cut into nibs. A simplified diagram of the system is shown in Figure 2.

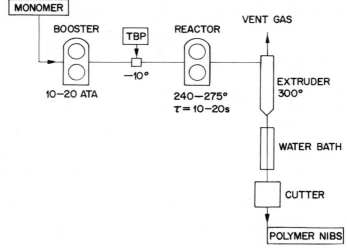

Figure 2. Melt-bulk polymerization process

The total residence time in the reaction zone is betwen 10 and 20 seconds. Initial experiments were carried out with reactor pumps hav-ing capacities of 1.2 ml and 20 ml per revolution, respectively. Figure 3 shows the initial laboratory setup of the reactor system.

The initiator's concentration determines the reaction rate, yield, and molecular weight of the product. (The limiting viscosity number, LVN, in benzyl alcohol at 150°C is used as a measure for the molecular weight.) An impression of these relationships can be gained from Figures 4 and 5.

In the systems tested, the heat losses were far too great for adiabatic conditions to be attained, and the reactors had to be heated externally. However; an increase in reactor size could result in a closer approach to the adiabatic state, leading perhaps to temperatures higher than 300°C.

Figure 3. Melt-bulk process: initial laboratory setup

INITIATOR (TBP) CONCENTRATION, %m ON LACTONE

□ 0.001
▲ 0.002
○ 0.005
△ 0.01
● 0.02
▽ 0.04
▼ 0.08

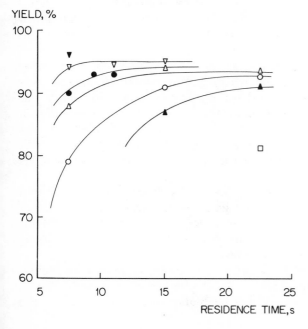

Figure 4. Melt-bulk process: yield-residence time relationship at various TBP concentrations

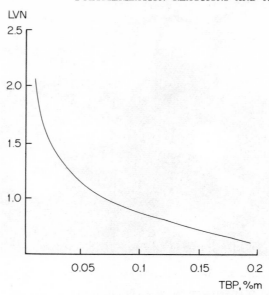

Figure 5. Melt-bulk process: MW control

At these temperatures, the reduction in yield caused by monomer decomposition might be considerable. This problem can be solved by using two reactors in series. By limiting the conversion in the first reactor, the temperature can be kept low enough. Conversion can be completed in the second reactor. Even if the temperature in the latter exceeds 300°C, decomposition of the monomer will be less of a problem than in a single reactor, as relatively much less monomer is exposed to the higher temperature.

To demonstrate this principle, we used a pipe attached to the outlet of the gear pump as the second reactor. The volume of this pipe was twice that of the gear pump, so that the residence time in the latter was reduced by two-thirds. To simulate the probable thermal situation in large-scale reactors, the temperature of the electrically heated pipe was raised to 350°C while the gear pump was maintained at 245°C. The advantages of this arrangement over a single reactor are shown in Table II.

The feasibility of the melt-bulk process has been demonstrated on a pilot-plant scale. In a 100 ml/rev gear pump (maximum capacity of 60 kg/h) conected to a 14-mm pipe reactor (volume of 150 ml), fiber-grade polymer (LVN 0.9-1.0) was produced in a continuous stable operation in yields of over 90% on intake.

We have not studied in detail the possibility of producing higher-molecular-weight material (LVN 1.5-2.0) for other applications (such as an engineering plastic). With the higher melt viscosities and the

Table II.   Melt-Bulk Process: Single *vs.* Two-Stage Reactor

	Temperature, °C		Residence time, s		Yield, %	LVN
	Pump	Pipe	Pump	Pipe		
Single	245	—	5	—	75–80	—
	245	—	12	—	92	1.0
	305	—	12	—	88.5	0.95
	355	—	10	—	86	0.85
Two-stage	245	275	5	16	93	0.99
	245	305	5	16	93	0.99
	245	355	5	16	93	0.95

lower initiator concentrations required (leading to lower reaction rates), we expect some difficulties in applying the bulk process to the manufacture of such material.

The next section deals with the development of our alternative process, in which the heat of polymerization is dissipated in a slurry system by the refluxing diluent. Although this process may appear not to be so economically attractive as the bulk route, it offers much more flexibility in the range of molecular weights that can be made.

Slurry Polymerization Process (9)

Polypivalolactone (PPL) is insoluble in most solvents at temperatures below 100°C. We therefore envisaged conducting the polymerization in an inert diluent from which the polymer precipitates, and in which the lactone monomer and initiator may or may not be soluble. The diluent should be capable of dissipating the heat of polymerization, and the morphology of the product (bulk density, powder flow, etc.) should be as good as possible. In addition, the reaction rates should be practical, and the polymer should fulfill product requirements such as thermal stability.

One approach we tried was a suspension process. The lactone monomer containing an initiator such as a phosphine was stirred at 60°-100°C in a long-chain hydrocarbon oil ("Shell Ondina" oil 33) in which it was essentially insoluble. Although polymer with a high bulk density was obtained in quantitative yields, the particles contained hydrocarbon oil that had to be removed by a hot organic wash (heptane) after recovery by filtration.

We also investigated the possibility of using a slurry system in which the monomer, dissolved in a refluxing hydrocarbon (hexane), was polymerized in the presence of a soluble initiator such as TBP. It turned out that the reaction rates had to be kept low on account of heat transfer problems, and the product was obtained as a very fine powder with a low bulk density. These factors consequently limited the maximum slurry concentration to 10-13 % w/w.

However, we found that products with high bulk densities (0.4-0.5) can easily be prepared by isolating the precipitated polymer at low conversion, and subsequently using it as initiator in a separate polymerization reaction. This concept of using "living" low-molecular-weight polypivalolactone (prepolymer) as a heterogeneous initiator led to the successful development of a slurry-polymerization process. Although metal salts such as potassium pivalate can also be used as heterogeneous initiators to give high-bulk density material, the thermal stability of the polymer during processing is at an unacceptably low level, as already described. A comparison of the various processes with diluent is given in Table III.

Table III. Polymerizations in Diluents

Type of process	Initiator	Limitations
Suspension	Phosphine	Hot Extraction Step Necessary
Slurry Homogeneous Initiator	Phosphine	Low Reaction Rate Low Bulk Density Low Maximum Slurry Concentration
Slurry Heterogeneous Initiator	Metal Salt	Unacceptable Thermal Stability
	Phosphine Prepolymer	

The structure of the prepolymer and its recommended characteristics are shown in Figure 6.

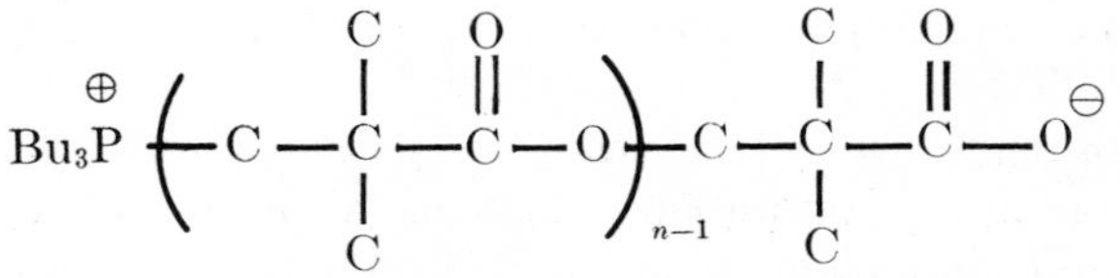

$$Bu_3P^{\oplus} \left(C - \overset{\overset{\displaystyle C}{|}}{\underset{\underset{\displaystyle C}{|}}{C}} - \overset{\overset{\displaystyle O}{\|}}{C} - O \right)_{n-1} C - \overset{\overset{\displaystyle C}{|}}{\underset{\underset{\displaystyle C}{|}}{C}} - \overset{\overset{\displaystyle O}{\|}}{C} - O^{\ominus}$$

Figure 6.   Slurry process: prepolymer structure

Molecular weight: 2-7 × 10³ (n=18-68)

Particle size: 10-177 μ

The prepolymer is prepared in a separate process step by adding the lactone to a solution of TBP in refluxing pentane. The particle size is governed by the stirring rate and the molecular weight by the lactone/phosphine molar ratio (Figure 7).

A feature of the slurry polymerization is that the reactivity of the carboxylate ion toward the lactone ring is much higher than that of the phosphine. It has been estimated that under the conditions of prepolymer preparation, the propagation rate is at least 600 times higher than the initiation rate. Another aspect of the system is the stability of the "living" polymer chain; although termination reactions can take place, prepolymer remains active almost indefinitely when stored at room temperature.

The main polymerization reaction is done by adding lactone to a stirred slurry of prepolymer in refluxing hexane. A chain-transfer agent

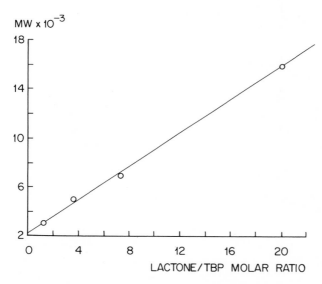

Figure 7.   Slurry process: prepolymer preparation

Table IV.   Slurry-Process Data

Prepolymer	: 0.5% w on lactone
Chain-transfer agent (EAA)	: 0.1–0.2% m on lactone
Diluent	: hexane
Reaction temperature	: 69 °C
Monomer addition	: 4 hours
Total reaction time	: 7 hours
Slurry conc.	: 43% w/w
Polymer recovery	: filtration
Yield	: 100%
Bulk density	: 0.5

such as ethyl acetoacetate (EAA) is added either to the monomer or to the solvent. Table IV presents preparative data for a typical polymerization.

The amount of chain-transfer agent required to obtain a given molecular weight depends to a certain extent on the monomer's purity. An indication of the relationship between LVN and the concentration of ethyl acetoactate that should be added to the solvent is given in Figure 8.

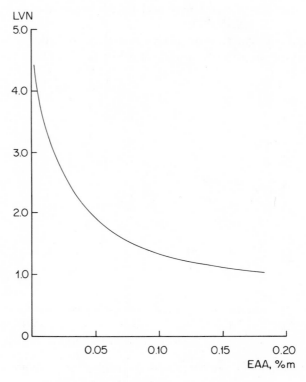

Figure 8. Slurry process: MW control

The morphology of the final powder depends mainly on that of the prepolymer; under optimal conditions, the product is a large-scale replica of the prepolymer particles. Figure 9 shows a typical product with a high bulk density (0.5) and a good flow caused by the spherical shape of the particles.

The slurry polymerization process with prepolymer has been scaled up from bench-scale (100 grams) to pilot-plant operation (0.5-ton batches) without major difficulties. An important aspect of this process is the absence of reactor fouling.

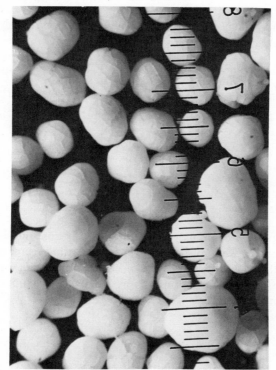

Figure 9. Slurry process: morphology of product (2.8 calibration spaces=100μ)

Properties of Polypivalolactone (PPL)

A major evaluation, including consumer testing, has been carried out on the polymer to determine its suitability for use as a textile fiber. Processing comprises melt spinning at 280°-290°C followed by stretching by usual procedures. In most mechanical properties, PPL is intermediate between polyethylene terephthalate (PET) and nylon-6. Its unique properties are its excellent elastic behavior, high elongation (6 g/denier at 80% elongation), and whiteness retention. Table V qualitatively compares the three fibers.

Table V.   Mechanical Properties [a]

	PPL	*Nylon-6*	*PET*
Tenacity	+	+	+
Stiffness	+	±	++
Knot strength	+	+	+
Elasticity	++	+	±
Color retention	++	±	++
Dyeability	±	++	+

[a] + = good; ± = fair; − = poor

As expected from its chemical structure PPL is highly resistant to chemicals, heat, and UV light, as shown in Table VI.

Table VI. Chemical Resistance [a]

	PPL	Nylon-6	PET
Alkali	+	+	±
Acid	+	±	+
Solvents	+	+	+
Heat	++	−	+
UV light	++	±	±

[a] + = good; ± = fair; − = poor

For plastic applications, PPL can be injection molded at 280°C. (Nucleating agent should be added.) Processing is easy because of good flow, high stability, and absence of carbonizing or crosslinking reactions. Its outstanding property as a potential engineering plastic is its high heat resistance (stable for 1000 hours at 200°C, or one year at 150°C).

The Future

The commercial future of polypivalolactone is difficult to predict at the moment. The problems of introducing a new plastic or fiber and competing against well-established materials are numerous. This is especially true in the present economic situation. We believe, however, that there is a place for polypivalolactone in the thermoplastics field and, as we have demonstrated, the technology and expertise now exists to prepare it on a commercial scale.

Acknowledgment

The author would like to acknowledge the contributions made to this work by numerous colleagues at the Koninklijke/Shell-Laboratorium: in particular D. P. Curotti, P. A. Desgurse, A. Klootwijk, and W. M. Wagner.

Literature Cited

1. Markevich, M. A., Pakhomeva, L. K., Enikolopyan, N. S., *Proc. Acad. Sci. USSR, Phys. Chem. Sect.* (1970) **187**, 499.

2. Jaacks, V., Mathes, N., *Makromol, Chem.* (1970) **131**, 295.
3. Jaacks, V., Mathes, N., *Makromol. Chem.* (1971) **142**, 209.
4. Letts, E. A., Collie, N., *Phil. Mag.* (1886) **22**, 183.
5. Denney, D. B., Kindsgrab, H. A., *J. Org. Chem.* (1963) **28**, 1133.
6. Netherlands Patent Application **7,002,062**.
7. Wiley, R. H., *J. Macromol. Sci. Chem.* (1970) **4, 1797**.
8. UK Patents **1,180,044** and **1,261,153**.
9. US Patents **3,578,700; 3,558,572; 3,471,456;** and **3,579,489**.

RECEIVED December 5, 1972.

12

Radiation-Induced Copolymerization of Hexafluoroacetone with α-Olefins

YONEHO TABATA, WATARU ITO, and KEICHI OSHIMA

University of Tokyo, Japan

YASUNOBU YAMAMOTO

Aishin Chemical Co., Kariya, Nagoya, Japan

Radiation induced copolymerization of hexafluoroacetone with α-olefins has been done over a broad temperature range. From these experiments, it was confirmed that ethylene can be copolymerized below its critical temperature to give an alternating copolymer. A radical mechanism is involved at relatively high temperatures; below $-10°C$, the mechanism has been confirmed to be ionic and may be cationic. Propylene can be copolymerized in a way similar to that of ethylene; however, the rate of copolymerization was much slower. Isobutylene did not copolymerize with hexafluoroacetone at $0°C$. A 1:2 adduct compound was obtained as the main product. At low temperatures, copolymerization proceeds to some extent.

Free-radical reactions of hexafluoroacetone have been studied (*1, 2, 3, 4*), and copolymerization of hexafluoroacetone with α-olefins at relatively high temperatures has been reported recently (*5, 6*). The copolymerization proceeds by a radical mechanism, giving a copolymer including small amounts of hexafluoroacetone. In this work, the copolymerization has been done at relatively low temperatures by gamma irradiation from a cobalt-60 source.

Experimental

Stainless-steel autoclaves of 10-ml capacity were used for this experiment (Figure 1). Commercial CF_3COCF_3 was used as the monomer. The monomer was 99.5% pure, with trace amounts of HF, HCl, and SO_2 as the impurities. Commercial-grade ethylene, propylene, and isobutylene were used as the comonomers.

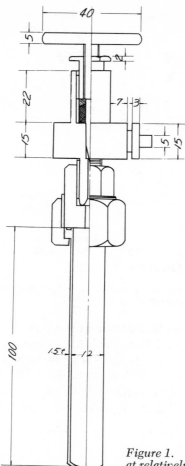

Figure 1.   Stainless-steel autoclave was used at relatively high temperatures (0° to 100°C) under pressure

The monomers were charged into the autoclave, and trace amounts of oxygen purged by the freeze-thaw technique.   Irradiation was carried out with gamma rays from a ^{60}Co source at various temperatures. $CHCl_3$, CH_2Cl_2, dimethylformamide (DMF), triethylamime (TEA), and p-benzoquinone were used as additives in the polymerization systems. Viscosity of the copolymers obtained was measured in Freon 113 at 35°C. Infrared spectra and x-ray diffraction patterns of the sample were also obtained.   NMR measurements of (proton and fluorine) were also done.

Results and Discussion

Copolymerization of Ethylene with Hexafluoroacetone. Conversion curves at −78°C for three different initial monomer compositions are shown in Figure 2. The conversion increases linearly with irradiation dose for higher concentrations of ethylene in the monomer mixture.

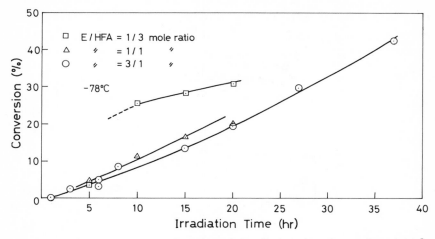

Figure 2. Conversions as a function of irradiation time for various initial monomer compositions at −78°C

The dependence of the copolymerization on temperature is shown in Figure 3. The polymerization rate decreases with a decreasing temperature from 0°C to −9°C; however, from −9°C to −100°C the rate increases with decreasing polymerization temperature. The existence of a minimum copolymerization rate around −10°C may be supported by previous experimental results (*1, 2, 3*) in which the copolymerization of

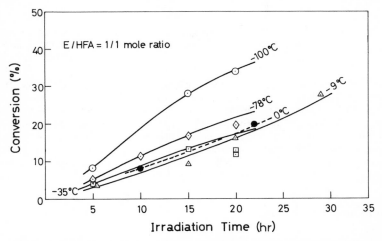

Figure 3. Conversions as a function of irradiation time at different temperatures for an equimolar monomer mixture

ethylene with hexafluoroacetone induced by radical initiators has been observed to be rapid at relatively high temperatures.

An Arrhenius plot of the copolymerization is shown in Figure 4. The result indicates that the polymerization mechanism is different between the higher-temperature and lower-temperature region. An activation energy of -0.8 kcal/mole was obtained for the region below $-9°C$. Above $-9°C$, the activation energy is about 2 kcal/mole.

Polymerization proceeds in a liquid and homogeneous phase. The negative activation energy of the copolymerization in the low-tempera-

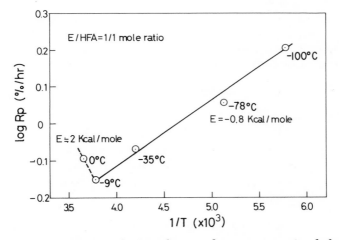

Figure 4. *Arrhenius plot of the copolymerization of ethylene-hexafluoroacetone*

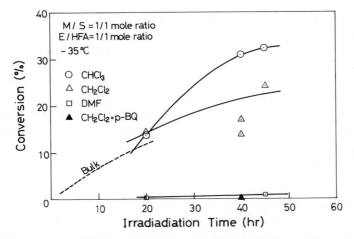

Figure 5. *Effect of additives on the copolymerization of ethylene hexafluoroacetone*

ture region suggests that the copolymerization proceeds *via* an ionic mechanism.

The effect of additives on the copolymerization was examined at $-35°C$. The results are shown in Figure 5. Under the conditions given, all of these additives are dissolved completely in the monomer mixture, and the copolymerization proceeds homogeneously during the early stages of the reaction. However, the reaction system gradually becomes inhomogeneous during the copolymerization. In the presence of TEA, only trace amounts of brownish, oily products were formed, and no polymerization was observed.

The results show that DMF inhibits the copolymerization completely. However, $CHCl_3$ and CH_2Cl_2 accelerate the copolymerization. It is known that halogenated hydrocarbons (RX) capture elctrons *via* dissociative electron attachment and stabilize monomer cation $M \cdot^+$.

$$M \longrightarrow \!\!\!\!\!\text{WWWW}\!\!\!\!\! \longrightarrow M \cdot^+ \; + \; e$$

$$RX + e \longrightarrow R \cdot \; + \; X^-$$

The lifetime of monomer cation should be lengthened, and a cationic polymerization is then favored to proceed under such circumstances.

However, amine compounds are known as cation scavengers, according to this scheme.

$$M \longrightarrow \!\!\!\!\!\text{WWWW}\!\!\!\!\! \longrightarrow M \cdot^+ \; + \; e$$

$$NR + M \cdot^+ \longrightarrow NRH^+ \; + \; M \cdot \; (-H)$$

Monomer cations are usually quenched effectively.

p-Benzoquinone inhibits radical polymerizations, and monomer cation radicals $M \cdot^+$ can also be attacked with the radical inhibitor. Although the inhibition mechanism of the cationic reaction by *p*-benzoquinone is not clear, the reactivity of the monomer cation is expected to be changed by its reaction with the radical inhibitor.

The fact that the copolymerization is inhibited by the addition of TEA indicates that the copolymerization proceeds *via* a cationic mechanism. The results given here show:

(a) Apparent negative activation of the polymerization at low temperatures.

(b) Enhancement of the copolymerization rate in the chlorinated hydrocarbon solvents.

(c) Inhibition of the copolymerization in the amine solvent.

Thus, a cationic polymerization is strongly supported for the copolymerization in the low-temperature region.

Infrared spectra of a copolymer and of a polyethylene are shown in Figure 6.   The spectrum of the copolymer is quite different from that of polyethylene.   Since the homopolymerization of hexafluoroacetone is difficult to achieve, the product obtained should be a copolymer.   Since the absorption arising from consecutive ethylene units is not observed,

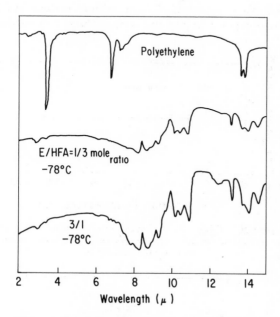

Figure 6.   Infrared spectra of copolymers and polyethylene

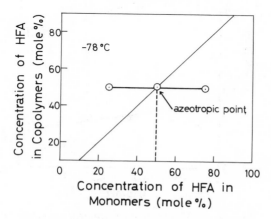

Figure 7.   Composition curve in the copolymerization of ethylene hexafluoroacetone

there is little possibility of sequences of ethylene units.   Therefore, the copolymer may be an alternating one.

A composition curve of the copolymerization is shown in Figure 7. The composition of one component in copolymers is exactly 50 mole % over a wide range of monomer concentration of the component in monomer mixtures.   These results strongly indicate that an alternating copolymerization takes place in the ethylene-hexafluoroacetone system.

Temperature dependency of the composition in copolymers was also

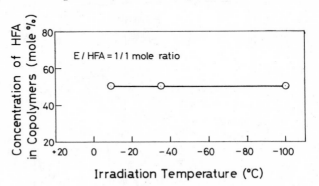

Figure 8.   Effect of polymerization temperature on the composition of copolymers for equimolar monomer mixture systems

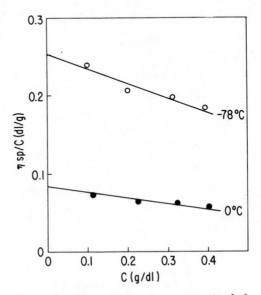

Figure 9.   Viscosity measurement of ethylene hexafluoroacetone copolymers in Freon 113 at 35°C

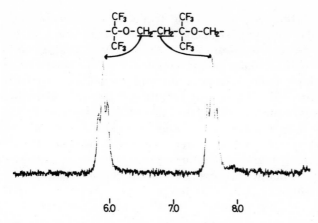

Figure 10.   Proton NMR spectrum of a copolymer obtained at −78°C; the measurement was made in Freon 113 at 50°C

examined, with the results shown in Figure 8.   The composition is not influenced at all by the copolymerization temperature, and an equimolar composition is always obtained.

Viscosities of the copolymers were measured in Freon 113 at 35°C (Figure 9).   For copolymers of hexafluoroacetone and ethylene, the intrinsic viscosity is much higher at −78°C than at 0°C.   A similar trend has been observed for the ternary system of ethylene–tetrafluoro-ethylene–hexafluoroacetone (7).

The lifetime of propagating polymer ion increases with decreasing temperature; therefore, the higher molecular weight at the lower polymerization temperature may be reasonably explained.   This is additional support for the ionic polymerization mechanism in the copolymerization at low temperatures.

The proton NMR spectrum of a copolymer obtained at −78°C is shown in Figure 10.   Two triplets of equal intensity centered at 5.93 and 5.62τ were observed. There is no other resonance for this sample.   On the basis of proton NMR spectra of model compounds (5), these two triplets may be assigned to two CH_2 groups, as shown below.

$$(CF_3)_2-\underset{\underset{\underset{H}{|}}{\underset{O-C-CH_3}{|}}}{C}-CH_2 \quad \text{and} \quad (CF_3)_2-\underset{\underset{O-CH_2}{|}}{C}-CHCH_3$$

7.55τ 5.63τ

The splitting of 7 cps in the triplet is a reasonable value for the vicinal H–H coupling, strongly indicating that ethylene units in the copolymer are isolated from each other. There are no continuous segments of ethylene units in the copolymer. For polymers obtained at relatively high temperatures, additional broadened resonances around 8.0 were observed. These may be attributed to methylene hydrogens in continuous polymeric units of ethylene and other structures.

It has been reported that hexafluoroacetone units are incorporated in copolymers randomly in lower concentrations than 50 mole % under high pressure at high temperatures (5). Our results suggest that deviation in the structure of copolymer from the alternating one increases with increasing polymerization temperature. At low temperatures, however, it is possible to obtain an alternating copolymer *via* a radical process.

A sharp singlet at 72.3 ppm from standard of CCl_3F was observed for copolymers obtained at $-78°C$. This is attributed to perfluoromethyl fluorine atoms in this copolymer:

$$-CH_2 - \underset{\underset{CF_3}{|}}{\overset{\overset{CF_3}{|}}{C}} - O - CH_2 -$$

An additional weak resonance at 77.2 ppm was observed for samples obtained at relatively high temperature. This may be attributed to fluorine atoms of:

$$-CH_2 - \underset{\underset{CF_3}{|}}{\overset{\overset{CF_3}{|}}{C}} - OH$$

The information from resonance patterns of both hydrogen and fluorine atoms completes the structure of copolymers.

An x-ray diffraction pattern of a copolymer obtained at $-78°C$ is shown in Figure 11. The copolymer is crystalline. Main diffraction peaks at 16.3°, 17.9°, and 27.8° for 2θ were observed. The regular alternating structure is the preferred one for crystalization.

Copolymerization of Propylene with Hexafluoroacetone. The conversion curves of the copolymerization are shown in Figure 12 as a function of irradiation time for equimolar monomer mixtures at 0°, $-35°$ and $-78°C$. The copolymerization rate is much slower than that of the copolymerization of ethylene and hexafluoroacetone. We have data a

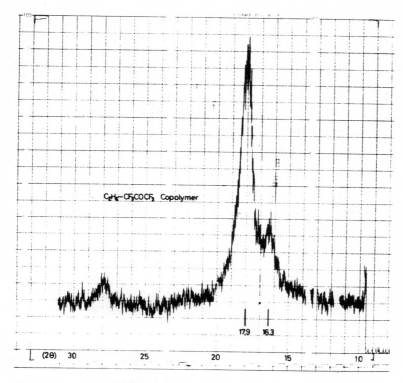

Figure 11.   X-ray diffraction of a copolymer of C_2H_4–$(CF_3)_2$ CO

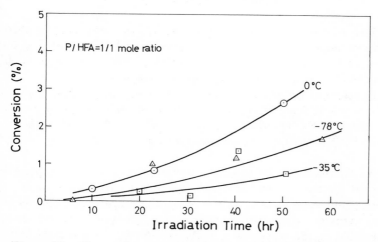

Figure 12.   Conversions as a function of irradiation time at different temperatures in the copolymerization of propylene-hexafluoroacetone

only three temperatures, but it is obvious that there is a similar tempera
ture dependency in the copolymerization to that in ethylene-hexafluoro
acetone copolymerization.   The rate of copolymerization increases with
increasing polymerization temperature above about $-38°C$, while the
rate increases with decreasing polymerization temperature below that
temperature.   These results suggest that a polymerization mechanism
similar to that of ethylene-hexafluoroacetone is operative in the copoly
merization of propylene and hexafluoroacetone.

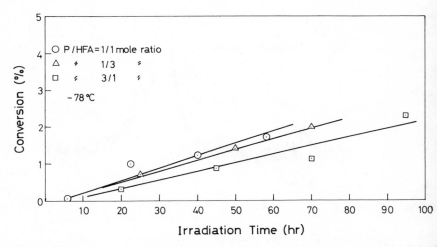

Figure 13.   Conversions as a function of irradiation time for various initial
monomer compositions at $-78°C$

The effect of initial composition of the monomer mixture the
copolymerization rate is shown in Figure 13.   A maximum rate was ob
served at the equimolar monomer mixture.   This fact suggests that an
alternative tendency exists in the copolymerization.

Reaction of Isobutylene with Hexafluoroacetone.   Conversion curves
of the reaction product appear in Figure 14.   The shape of the conver
sion curve at $0°C$ is different from curves at $-35°C$ and $-78°C$.   The
effect of initial monomer concentration on the reaction is shown in Figure
15.

A different type of reaction from those at lower temperatures is
expected for the reaction at $0°C$.   According to infrared spectra of the
products obtained at $0°C$ and $-78°C$, there are major differences be
tween the two.   Furthermore, the products are quite different from each
other—that is, the former is a white powder product and the latter is a
rubberlike material.   The former seems to be a low-molecular-weight
adduct, and the latter is certainly a polymeric product.

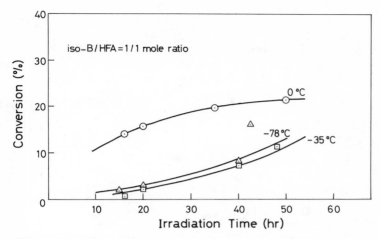

Figure 14. Conversions as a function of irradiation time at three different temperatures in the copolymerization of isobutylene-hexafluoroacetone

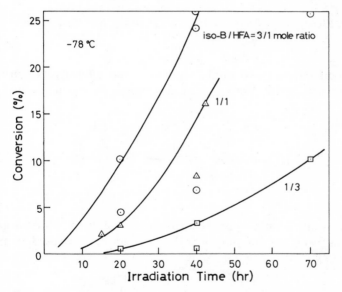

Figure 15. Conversions as a function of irradiation time in the copolymerization of isobutylene hexafluoroacetone for different initial monomer compositions at −78°C.

The proton NMR spectrum of the product obtained at 0°C is shown in Figure 16. Main resonances are a doublet 7.90τ and two singlets at 6.52τ and 4.25τ. Their intensity ratio was 3:2:1. Weak resonances were observed at 7.56τ, 6.77τ and 4.64τ. The fluorine NMR spectrum of the

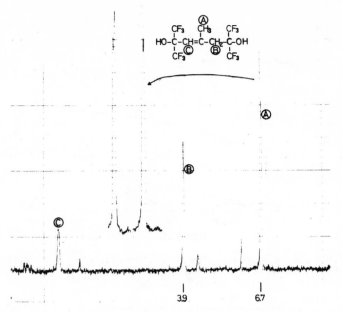

Figure 16. Proton NMR spectrum of the product obtained at 0°C; the measurement was done in pyridine at 50°C

same product shows two singlets having equal intensity (Figure 17).

From the NMR spectra, it is concluded the main product is this adduct:

$$\underset{\underset{CF_3}{|}}{\overset{\overset{CF_3}{|}}{HO-C-CH}} = \underset{\underset{CH_3}{|}}{\overset{\overset{CH_3}{|}}{C}} -CH_2-\underset{\underset{CF_3}{|}}{\overset{\overset{CF_3}{|}}{C}}-OH$$

This result is consistent with the elemental analysis of the product. There has been no (or only a little) 1:1 adduct observed in previous work (*2*).

$$CH_2 = \underset{\underset{CH_3}{|}}{C} -CH_2-\underset{\underset{CF_3}{|}}{\overset{\overset{CF_3}{|}}{C}}-OH$$

The melting point of the product was 145°C, corresponding to that of the former product.

From these results, it can be concluded that the main reaction at 0°C is not a copolymerization but an adduct formation of two monomers in the ratio of 1:2. The main reaction at −78°C may be either co-polymerization of two monomers or both copolymerization and homo-polymerization of isobutylene.

Small amounts of hexafluoroacetone are incorporated in the polymer chain. An ionic mechanism is probably operative in the polymeriza-tion at −78°C.

Polymerization Mechanism

The electronic arrangement of the carbonyl function in fluoroketones more closely resembles the double bond of weakly polarized olefins than it does the carbonyl group in normal hydrocarbon ketones, and fluoro-ketones may therefore undergo free-radical reactions more readily than their hydrocarbon counterparts (3).

According to studies on the reactivity of hexafluoroacetone, these carbonyl groups are very reactive to radicals. There are two possi-bilities for the addition of R radicals to fluorocarbonyl groups:

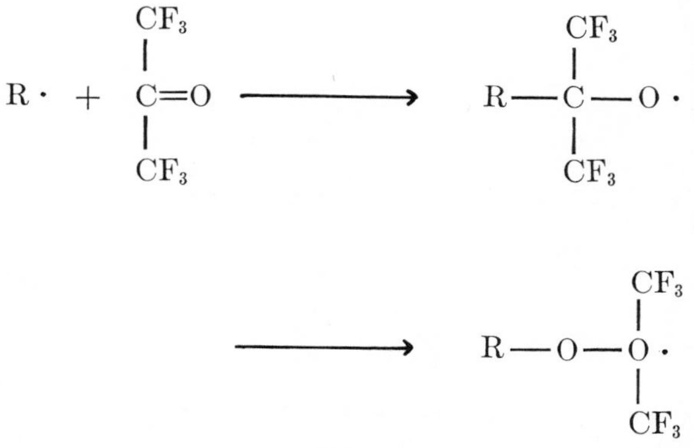

$$
R \cdot \; + \; \overset{\displaystyle CF_3}{\underset{\displaystyle CF_3}{C=O}} \quad \longrightarrow \quad R\!-\!\overset{\displaystyle CF_3}{\underset{\displaystyle CF_3}{C}}\!-\!O \cdot \qquad (A)
$$

$$
\longrightarrow \quad R\!-\!O\!-\!\overset{\displaystyle CF_3}{\underset{\displaystyle CF_3}{O}} \cdot \qquad (B)
$$

Both possibilities give oxy and ether compounds.

It is believed that the intermediate B should be more stable than A. The intermediate stability and steric hindrance suggest that addition to oxygen should be the preferred mode of addition of free radicals to fluoroketones:

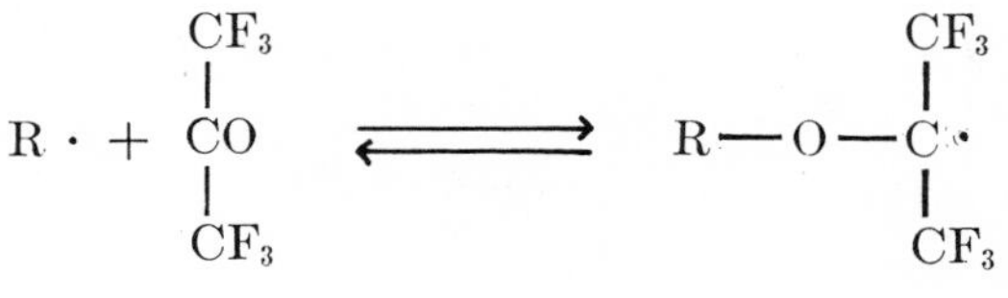

$$
R \cdot \; + \; \overset{\displaystyle CF_3}{\underset{\displaystyle CF_3}{CO}} \quad \rightleftarrows \quad R\!-\!O\!-\!\overset{\displaystyle CF_3}{\underset{\displaystyle CF_3}{C}} \cdot
$$

In our polymerization at relatively low temperatures ($0°$ and $-9°$), radical reactions may take place to give polymers. The main polymerization process is considered to be:

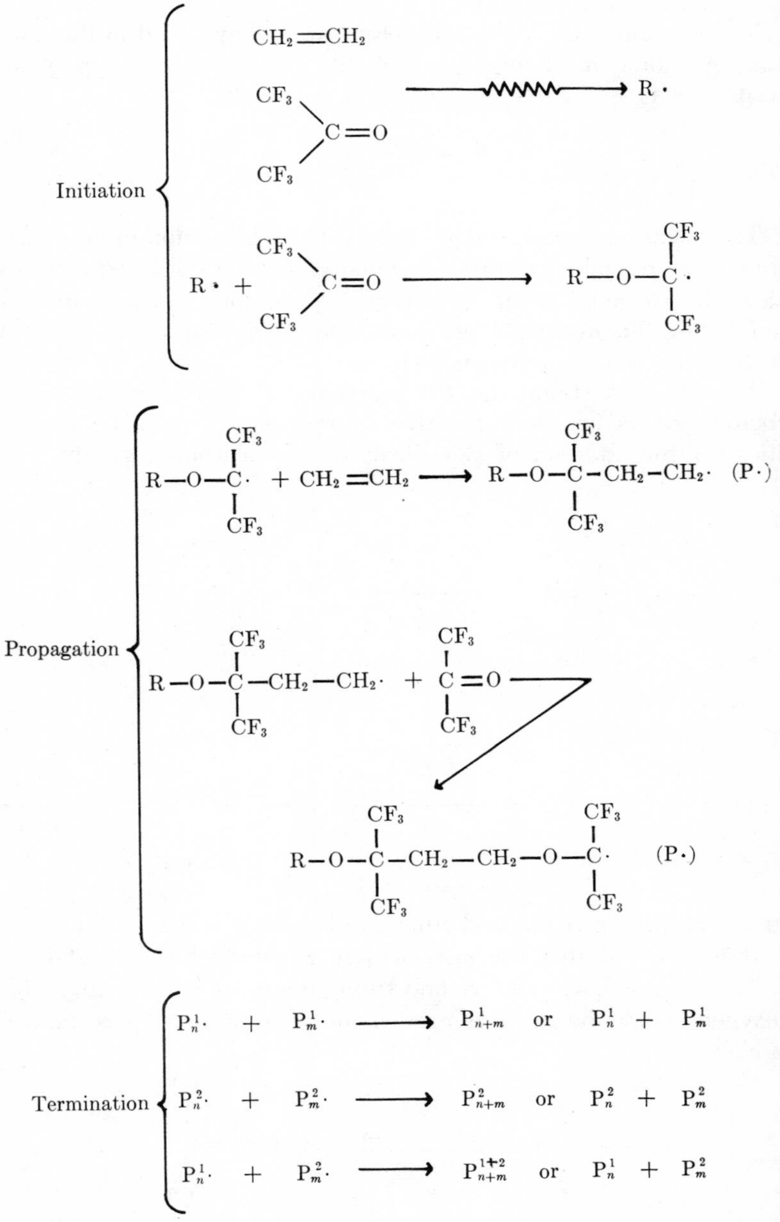

Initiation

$$CH_2=CH_2$$

$$\begin{array}{c} CF_3 \\ \diagdown \\ C=O \\ \diagup \\ CF_3 \end{array} \xrightarrow{\hspace{2cm}\text{\Large\char"223F}\hspace{1cm}} R\cdot$$

$$R\cdot + \begin{array}{c} CF_3 \\ \diagdown \\ C=O \\ \diagup \\ CF_3 \end{array} \longrightarrow R-O-\overset{\displaystyle CF_3}{\underset{\displaystyle CF_3}{\overset{|}{\underset{|}{C}}}}\cdot$$

Propagation

$$R-O-\overset{\displaystyle CF_3}{\underset{\displaystyle CF_3}{\overset{|}{\underset{|}{C}}}}\cdot + CH_2=CH_2 \longrightarrow R-O-\overset{\displaystyle CF_3}{\underset{\displaystyle CF_3}{\overset{|}{\underset{|}{C}}}}-CH_2-CH_2\cdot \quad (P\cdot)$$

$$R-O-\overset{\displaystyle CF_3}{\underset{\displaystyle CF_3}{\overset{|}{\underset{|}{C}}}}-CH_2-CH_2\cdot + \overset{\displaystyle CF_3}{\underset{\displaystyle CF_3}{\overset{|}{\underset{|}{C}}}}=O \longrightarrow$$

$$R-O-\overset{\displaystyle CF_3}{\underset{\displaystyle CF_3}{\overset{|}{\underset{|}{C}}}}-CH_2-CH_2-O-\overset{\displaystyle CF_3}{\underset{\displaystyle CF_3}{\overset{|}{\underset{|}{C}}}}\cdot \quad (P\cdot)$$

Termination

$$P_n^1\cdot + P_m^1\cdot \longrightarrow P_{n+m}^1 \quad \text{or} \quad P_n^1 + P_m^1$$

$$P_n^2\cdot + P_m^2\cdot \longrightarrow P_{n+m}^2 \quad \text{or} \quad P_n^2 + P_m^2$$

$$P_n^1\cdot + P_m^2\cdot \longrightarrow P_{n+m}^{1+2} \quad \text{or} \quad P_n^1 + P_m^2$$

Homopolymerization of ethylene is extremely low under the conditions mentioned above (8). Therefore, by these polymerization mechanism, an alternating copolymerization has been deduced. This mechanism is very different from that for copolymerizations carried out at relatively high temperatures under high pressures. In the latter case, small amounts of hexafluoroacetone were incorporated in copolymers. Furthermore, at relatively high temperatures, (A) type of reaction may increase and a hydrogen-transfer type of reaction may occur, following the addition reaction, to give hydroxyl groups.

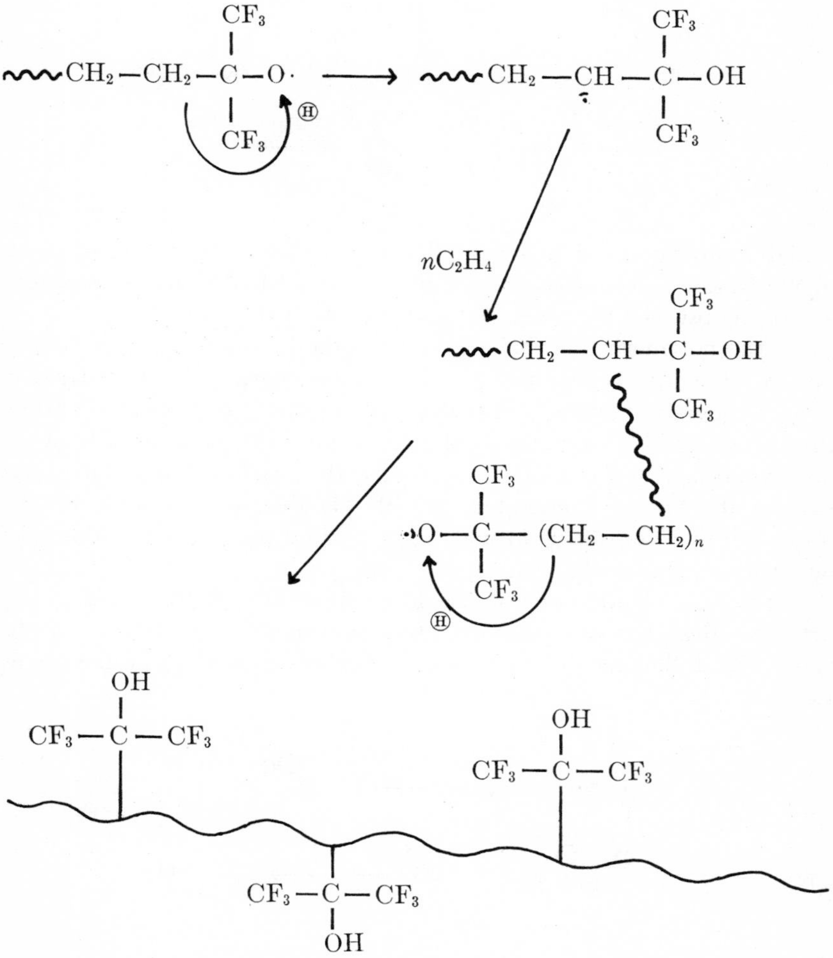

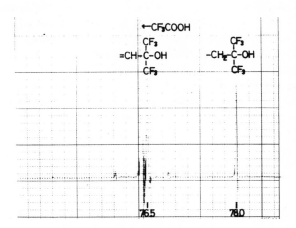

Figure 17.   Fluorine NMR spectrum of the product; the measurement was made in dioxane at 80°C

A small contribution of this type of reaction cannot be ruled out in our polymerization system because small amounts of hydroxyl in copolymers have been observed by infrared measurements of the samples.

In the polymerization below $-35°C$, another kind of polymerization must be at work because negative activation energy has been observed in the temperature region.   Usually, the reactivity of radicals (particularly in radical chain reactions) should be very low or negligible at low temperatures.   On the contrary, however, the reaction takes place much faster at the lowest temperature ($-100°C$) than do reactions at other temperatures.   This result strongly suggests that the polymerization takes place *via* an ionic mechanism.

The effect of additives on the low-temperature polymerization supports the idea that the polymerization proceeds *via* a cationic mechanism.   The following polymerization mechanism may be proposed tentatively:

Propagation

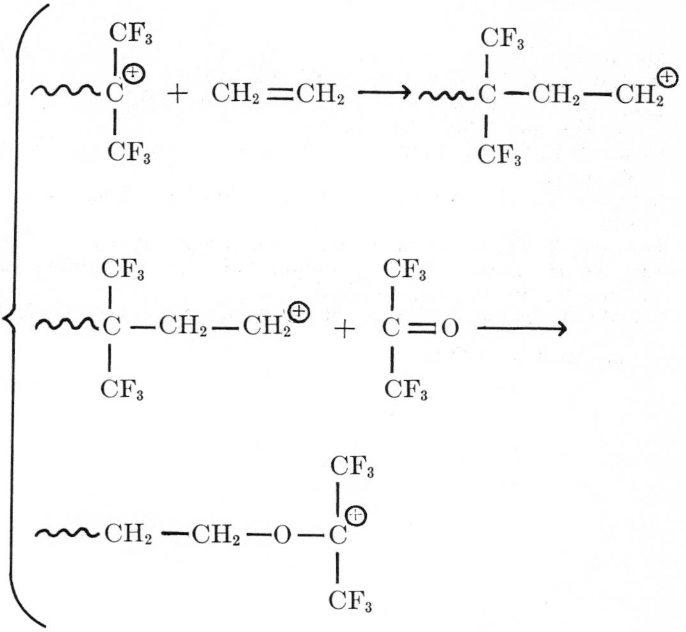

$$
\underset{\sim}{\overset{\overset{\displaystyle CF_3}{\vert}}{\underset{\underset{\displaystyle CF_3}{\vert}}{C}}}{}^{\oplus} + \ CH_2{=}CH_2 \ \longrightarrow \ \underset{\sim}{\overset{\overset{\displaystyle CF_3}{\vert}}{\underset{\underset{\displaystyle CF_3}{\vert}}{C}}}{-}CH_2{-}CH_2{}^{\oplus}
$$

$$
\underset{\sim}{\overset{\overset{\displaystyle CF_3}{\vert}}{\underset{\underset{\displaystyle CF_3}{\vert}}{C}}}{-}CH_2{-}CH_2{}^{\oplus} + \ \overset{\overset{\displaystyle CF_3}{\vert}}{\underset{\underset{\displaystyle CF_3}{\vert}}{C}}{=}O \ \longrightarrow
$$

$$
\underset{\sim}{}CH_2{-}CH_2{-}O{-}\overset{\overset{\displaystyle CF_3}{\vert}}{\underset{\underset{\displaystyle CF_3}{\vert}}{C}}{}^{\oplus}
$$

Termination

$$
\underset{\sim}{}CH_2{-}CH_2{}^{\oplus} + \ M^- \ or \ e \ \underset{\sim}{}CH_2{-}CH_2{\cdot} \ + \ M
$$
$$
\longrightarrow
$$

$$
\underset{\sim}{}O{-}\overset{\overset{\displaystyle CF_3}{\vert}}{\underset{\underset{\displaystyle CF_3}{\vert}}{C}}{}^{\oplus} + \ M^- \ or \ e \ \underset{\sim}{}O{-}\overset{\overset{\displaystyle CF_3}{\vert}}{\underset{\underset{\displaystyle CF_3}{\vert}}{C}}{\cdot} \ + \ M
$$

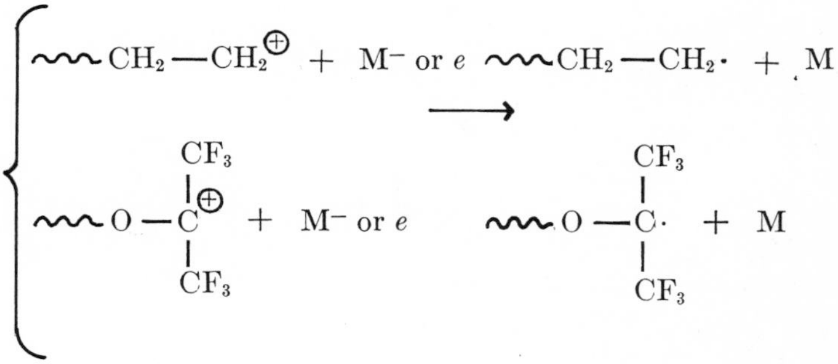

The formation of a 1:1 charge transfer complex or molecular complex and its contribution to the copolymerization can be easily expected. However, there is no direct evidence at present. The study is continuing to clarify the polymerization mechanism.

Acknowledgment

We wish to express our thanks to Mr. Takayama for his technical assistance, and to K. Ishigure for both his measurements of NMR spectra and his discussion of the experimental results.

Literature Cited

1. Middleton, W. J., U.S. Patents **224,211** (1962) and **293,126** (1963); Howard, E. G., U.S. Patent **363,262** (1964).
2. Litt, M. H., Schmitt, G. J., U.S. Patents **3,324,187** (1967), **3,358,002** (1967), and **3,395,235** (1968).
3. Howard, E. G., Sargeant, P. B., Krespan, C. G., *J. Am. Chem. Soc.* (1967) **89,** 1422.
4. Krespan, C. G., Middleton, W. J., *Fluorine Chem. Rev.* (1967) **1(1),** 145.
5. Howard, E. G., Sargeant, P. B., *J. Macromol. Sci.-Chem.* (1967) **A1,** 1011.
6. Litt, M. H., Bouer, F. W., *J. Polym. Sci., Part C.* (1965) **16,** 1551.
7. Tabata, Y., Ito, W., unpublished data.
8. Tabata, Y., Hara, A., Sobue, H., *J. Polym. Sci.* (1964) **A2,** 4077.

RECEIVED April 13, 1972.

13

Donor-Acceptor Complexes in Copolymerization

XXXI. Alternating Copolymer Graft Copolymers
VI. Synthesis in the Absence and Presence of Complexing Agents

NORMAN G. GAYLORD

Gaylord Research Institute Inc., New Providence, N.J. 07974

Graft copolymers containing alternating copolymer branches were prepared by copolymerizing monomers that form alternating copolymers under conditions necessary for formation of the specific alternating copolymer—i.e., with or without a complexing agent, in either case with or without a radical catalyst, and in the presence of polymers containing labile or active hydrogen atoms. Alternating graft copolymers of styrene–maleic anhydride, isoprene-maleic anhydride, styrene-methyl methacrylate, and styrene-acrylonitrile were prepared in bulk, solution, or aqueous medium on one or more of these polymers: polystyrene, poly-(butyl acrylate), high- and low-density polyethylene, atactic and isotactic polypropylene, EPR, nitrile rubber, cis-1,4-polybutadiene, ABS, PVC, and cellulose.

The copolymerization of a strong donor monomer and a strong acceptor monomer—e.g., styrene and maleic anhydride—leads to the formation of an equimolar alternating copolymer over a wide range of monomer feed ratios. Although the polymerization may occur spontaneously at elevated temperatures, the addition of a free-radical precursor increases the yield of alternating copolymer, independent of monomer charge (*1*).

The copolymerization of a donor monomer with a weaker acceptor monomer, such as methyl methacrylate or acrylonitrile, in the presence of a complexing agent (a metal halide or an organoaluminum halide) yields an equimolar, alternating copolymer, regardless of initial monomer charge and with or without free-radical precursor (*1*).

Polymerization of Donor-Acceptor Comonomer Complexes

The formation of alternating copolymer is attributed to the homopolymerization of a comonomer charge-transfer complex. The latter is formed spontaneously, subject to equilibrium considerations, with the interaction of a strong donor monomer and a strong acceptor monomer.

$$D + A \rightleftharpoons [D \longrightarrow A]$$

The formation of the charge-transfer complex is enhanced by complexing an acceptor monomer with a metal halide, as a consequence of the increased ability of the complexed monomer to accept electrons.

$$A + MX \longrightarrow A \ldots MX$$

$$D + A \ldots MX \rightleftharpoons [D \longrightarrow A \ldots MX]$$

Spontaneous polymerization occurs when the equilibrium concentration of the complex is high enough. Lower temperatures favor complex formation, while polymerization may require thermal or catalytic activation. Since the polymerization with and without the radical catalyst follows the same course, the comonomer complex is considered to be the polymerizable species in either case.

Although the formation of the ground-state comonomer charge-transfer complex occurs spontaneously, it has been suggested (2, 13) that polymerization involves the excited-state complex (exciplex).

$$[D \longrightarrow A] \rightleftharpoons [D^+ . {}^-A] \longrightarrow -(DA)_x-$$

$$[D \longrightarrow A \ldots MX] \rightleftharpoons [D^+ . {}^-A \ldots MX] \longrightarrow -(DA)_x- + MX$$

The exciplex is apparently generated spontaneously when the concentration of ground-state complexes is high enough, probably because of collision. The effect of thermal activation is to increase the number of collisions.

The proposed polymerization mechanism (1) involves initiation by a head-to-head reaction of two excited complexes; during the reaction, hydrogen abstraction and transfer occurs through a six-membered cyclic transition state to generate a polymer chain with an attached terminal complex. The propagation reaction is head-to-tail and involves the addition of excited complexes to the chain end through hydrogen transfer *via* a six-membered cyclic transition state to regenerate the complex chain end.

The role of the free-radical initiator was earlier considered to involve hydrogen abstraction (1). However, decomposition of the free-radical precursor may also result in excitation of the ground-state complexes (2).

Chlorine is virtually absent in the copolymer produced in the azobisisobutyronitrile (AIBN) catalyzed copolymerization of styrene and maleic anhydride in the presence of chloroform or carbon tetrachloride (3, 4), or of p-dioxene and maleic anhydride in the presence of acrylonitrile in chloroform (5). This absence indicates that trichloromethyl radicals generated by the reaction of the chlorinated hydrocarbons with the radicals from AIBN are not incorporated into the polymer chain. Similarly, there is little or no chlorine in the alternating copolymer that is formed in the copolymerization of styrene and methyl methacrylate in the presence of ethylaluminum sesquichloride (EASC) in the presence of chloroform and carbon tetrachloride, and with or without a peroxide initiator (6).

The failure to incorporate moieties arising from radical attack on the solvent into the alternating copolymers, coupled with the virtual absence of catalyst residues in the copolymer when the copolymerization of styrene and maleic anhydride is initiated by AIBN (3, 4), indicates that radical species may initiate the polymerization of comonomer charge-transfer complexes, but they are not incorporated into the polymer chain.

The absence of chlorine in the copolymers, as well as the formation of high-molecular-weight copolymers in reactions carried out in chloroform or carbon tetrachloride, indicates that the propagating chains do not participate in chain-transfer reactions that result in abstraction of hydrogen or chlorine atoms. High-molecular-weight alternating copolymers are also obtained in the radical-catalyzed copolymerization of propylene and maleic anhydride in liquid propylene (7), and in the spontaneous copolymerization of propylene and other α-olefins with acrylonitrile in the presence of an organoaluminum halide (8). This is further evidence that the propagating chain in the homopolymerization of comonomer charge-transfer complexes does not abstract labile hydrogen atoms even from α-olefins, which usually undergo degradative chain transfer in the presence of free radicals.

Grafting of Alternating Copolymers

Despite the absence of both chain transfer by the propagating chain and the initiating radical in the copolymer chain, graft copolymers containing alternating copolymer branches can be prepared by copolymerizing monomers that form alternating copolymers in the presence of polymers containing labile or active hydrogen atoms—e.g., tertiary or allylic carbon atoms and aldehyde groups (9).

One proposal (1) holds that the propagating chain end in the homopolymerization of a comonomer complex is a complex, and that bimolecular termination involves polymer-polymer or polymer-complex interaction through carbene formation. It is reasonable to propose that when the comonomers are polymerized in the presence of a polymer, grafting occurs by termination of a propagating alternating copolymer chain on a labile C–H of the substrate polymer, *i.e.*, by insertion of the polymeric carbene into the labile C–H of the substrate.

$$
\begin{array}{ccc}
\underset{\text{XCY}}{\overset{\text{H}}{\text{HC}^-}} \quad \underset{\text{HCH}}{\overset{\text{H}}{^+\text{C}-\text{R}}} & & \underset{\overset{|}{\text{H}}}{\underset{\text{XCY}}{\overset{\text{H}}{\text{HC}}}} \quad \underset{\text{HCH}}{\overset{\text{H}}{\text{C}-\text{R}}} \\
(\text{H}) & \longrightarrow & \\
\underset{\underset{\text{H}}{\text{R}-\text{C}^+}}{\text{P}-\text{C}} \quad \underset{\underset{\text{H}}{-\text{CH}}}{\text{YCX}} & & \underset{\underset{\text{H}}{\text{R}-\text{C}}}{\text{P}-\dot{\text{C}}} \quad \underset{\underset{\text{H}}{\text{CH}}}{\text{YCX}}
\end{array} \quad \overset{\sim\overset{|}{\text{C}}\sim}{\underset{\text{H}}{}} \longrightarrow
$$

$$
\begin{array}{c}
\sim\overset{|}{\underset{|}{\text{C}}}\sim \\
\text{P}\!\!-\!\!\overset{|}{\underset{|}{\text{CH}}} \\
\text{R}\!\!-\!\!\overset{|}{\underset{|}{\text{CH}}} \\
\overset{|}{\underset{|}{\text{HCH}}} \\
\overset{|}{\underset{|}{\text{YCX}}} \\
\overset{|}{\underset{|}{\text{HCH}}} \\
\text{R}\!\!-\!\!\overset{|}{\underset{|}{\text{CH}}} \\
\overset{|}{\underset{|}{\text{HCH}}} \\
\overset{|}{\underset{\text{H}}{\text{YCX}}}
\end{array}
$$

Grafting in the Absence of Complexing Agents

The grafting of an alternating copolymer on a substrate polymer occurs when the alternating copolymer is prepared under conditions normally used without the substrate polymer. When comonomers that are subject to spontaneous or thermal initiation are polymerized in the presence of a suitable polymer, graft copolymers are formed despite the absence of a radical catalyst (*10*).

Table I.   Graft Copolymerization of Poly (styrene-alt-maleic anhydride)
onto Polystyrene

S-MAnh/PS wt ratio	0.5	1.0	1.0	9.0
Polystyrene (PS), grams	5.0	2.0	20.0	1.0
M_v	10,800	10,800	40,500	40,500
S/MAnh (1/1 m), grams	1.29/	1.03/	10.3/	4.63/
	1.21	0.97	9.7	4.36
Benzene, ml	7.5	3.0	30	0
Temp/Time, °C/hr	60/3	60/3	80/1	60/3
Monomer conversion, %	26.4	40.8	70.8	45.8
Fractionation [a]				
Polystyrene, %	0.7	0.2	23.7	0.8
Graft copolymers				
Benzene soluble, %	91.9	73.8	43.8	0.0
PS/S-MAnh, mole %	83/17	87/13	94/6	—
Benzene insoluble, %	7.4	25.9	32.5	99.2
PS/S-MAnh, mole %	7/93	0/100	8/92	3/97

[a] Precipitation in 1:2 benzene-petroleum ether from MEK solution.

Typical results in the graft polymerization of poly(styrene-alt-maleic anhydride) onto polystyrene in solution without a radical catalyst are shown in Table I.   Analogous results are obtained in the presence of a radical catalyst.

The molecular weight of an alternating copolymer depends on the concentration of polymerizable species, the comonomer charge-transfer complex, which in turn depends on the reaction temperature.   At elevated temperatures, the copolymerization rate is rapid and the molecular weight of the alternating copolymer is low.   When the polymerization is carried out under those conditions in the presence of a polymer, the grafted alternating copolymer is present as multiple short branches.

Spontaneous copolymerization of styrene and maleic anhydride in the presence of a molten polymer or a bulk polymer undergoing deformation at elevated temperatures is a rapid and convenient route for carboxylating polymers.   The reaction is carried out on the bulk polymer at 120° to 200°C. (depending upon the softening or melting point of the polymer) by injecting an equimolar solution of maleic anhydride in styrene into the molten polymer (*11, 12*).

The reaction is done in equipment such as a Brabender Plasticorder or extruder; it can be a batch or a continuous process.   Maleic anhydride is dissolved in styrene at 60°C, and the solution (still at 60°C) is injected into the molten polymer which is being fluxed or mixed at the desired temperature.   Since the polymerization and grafting reactions are just about instantaneous, no reaction time is needed after monomer addition is finished.

When the grafting reaction is done in the Brabender Plasticorder, the polymer is fluxed for two minutes. The monomer mixture is continuously injected into the molten mass over one minute, and the reaction mixture then discharged from the Plasticorder.

When the reaction is done in an extruder, the rate of monomer addition, generally through a vent into the barrel, is adjusted depending upon the rate of throughput of the molten polymer. The carboxylated polymer may be extruded as ribbon, which is then formed into pellets or as a fiber or film.

Table II shows typical results obtained in the carboxylation of low-density polyethylene in the Plasticorder at 150°C. Extraction of the reaction product with acetone at 25°C for 24 hours removes unreacted monomers and ungrafted styrene-maleic anhydride copolymer.

Table II.   Carboxylation of Low-Density Polyethylene at 150°C [a]

Monomers	S/MAnh	S/AA	S/AA	MAnh
Weight, grams	2.7/6.2	2.7/1.9	2.7/3.8	2.6
Mole ratio	1/1	1/1	1/2	—
LDPE, grams	40.0	40.0	40.0	40.0
Recovery, grams	44.9	41.9	43.2	41.5
Monomer Conversion, %	92.4	41.3	49.2	57.7
Acetone-Extraction				
Soluble, %	0.4	0.6	0.8	1.8
Insoluble, %	99.6	99.4	99.2	98.2
Acid Value, mg KOH/g				
Before Extraction	35.4	11.2	10.9	19.7
	(70.8)			(39.4)
After Extraction	31.9	6.2	5.6	2.2
	(63.8)			(4.4)

[a] Two min in Brabender Plasticorder.

The acid value is obtained by refluxing the product with xylene to convert carboxylic acid groups (formed by exposure of the product to air) into anhydride groups, followed by refluxing with anhydrous methanolic KOH. The cooled mixture is then titrated with 2-propanolic HCl using phenolphthalein as the indicator. The acid value shown in Table II is the actual titration value. However, since methanol converts the anhydride to the half ester, the true acid value, shown in parentheses, is double the titration value.

The effectiveness of the reaction with styrene-maleic anhydride is clear when the results are compared with those obtained—under the same conditions without a radical catalyst—using a 1:1 molar mixture of styrene and acrylic acid, a 1:2 molar mixture of styrene and acrylic

Table III. Carboxylation of High-Density Polyethylene [a]

Temperature, °C	150	170
S/MAnh (1/1 m), grams	2.7/2.6	2.7/2.6
HDPE, grams	40.0	40.0
Recovery, grams	44.1	43.4
Monomer Conversion, %	77.4	64.2
Acetone Extraction		
Soluble, %	3.2	3.2
Insoluble, %	96.8	96.8
Acid Value, mg KOH/g		
Before Extraction	32.2 (64.4)	34.0 (68.0)
After Extraction	20.0 (40.0)	23.7 (47.4)

[a] Two min in Brabender Plasticorder.

Table IV. Carboxylation of Atactic Polypropylene at 150°C [a]

S/MAnh (1/1 m), grams	1.4/1.3	2.7/2.6
APP, grams	40.0	40.0
Acetone Extraction		
Soluble, %	9.8	5.8
Insoluble, %	91.8	94.2
Acid Value, mg KOH/g		
Before Extraction	24.3 (48.6)	59.6 (119.2)
After Extraction	21.3 (42.6)	32.5 (65.0)

[a] Two min in Brabender Plasticorder.

acid (to give the same number of carboxyl groups as are available in maleic anhydride), and maleic anhydride alone (Table II).

The monomer conversion using the mixture of styrene and maleic anhydride is much higher than in the other cases. This value does not truly represent monomer conversion but, instead, the amount of monomer not lost during treatment. Thus, both reacted monomers and non-volatilized monomers are present in the reported value. The values obtained with maleic anhydride alone represent a large amount of un-reacted maleic anhydride, as indicated by the amount extracted and the acid value of the polymer after extraction. In any event, the recovery with styrene-maleic anhydride is considerably higher than in all the other cases.

The acetone-soluble fraction representing unreacted and reacted monomers that did not react with the substrate polymer is also far smaller with the styrene-maleic anhydride mixture.

The acid values, both before and after extraction with acetone, are at least five times higher with styrene-maleic anhydride than with any of the other mixtures. The differences are even more dramatic when the acid values are doubled for styrene-maleic anhydride.

The results obtained by carboxylating high-density polyethylene are shown in Table III. The tertiary carbon atoms at branch points are grafting sites. The high acid values may also indicate that the grafted styrene-maleic anhydride copolymer chains are of higher molecular weight than in the case of low-density polyethylene.

The presence of a great number of grafting sites is shown in the carboxylation of atactic polypropylene at 150°C (Table IV). Isotactic polypropylene requires a temperature of 170°C for effective carboxylation (Table V). The relatively greater amount of extractable material in the case of atactic polypropylene indicate the acetone solubility of the polypropylene or the lightly grafted polymer, or both.

Comparison of the carboxylation procedures using acrylic acid in lieu of maleic anhydride is even more striking with isotactic polypropylene (Table V) than with polyethylene.

Table V. Carboxylation of Isotactic Polypropylene at 170°C [a]

Monomers	S/MAnh	S/AA	S/AA	MAnh
Weight, grams	2.7/2.6	2.7/1.9	2.7/3.8	2.6
Mole Ratio	1/1	1/1	1/2	—
IPP, grams	40.0	40.0	40.0	40.0
Recovery, grams	44.2	42.4	42.8	41.4
Monomer Conversion, %	79.2	52.2	43.1	53.8
Acetone Extraction				
Soluble, %	1.4	2.2	3.2	2.6
Insoluble, %	98.6	97.8	96.8	97.4
Acid Value, mg KOH/g				
Before Extraction	27.9	7.4	6.8	24.1
	(55.8)			(48.2)
After Extraction	15.9	1.3	2.1	1.9
	(31.8)			(3.8)

[a] Two min in Brabender Plasticorder.

Table VI. Carboxylation of Ethylene-Propylene Rubber [a]

Temperature, °C	100	150	170
S/MAnh (1/1 m), grams	3.6/3.4	3.6/3.4	3.6/3.4
EPR, grams	35.0	35.0	35.0
Acetone Extraction			
Soluble, %	8.0	2.5	3.8
Insoluble, %	92.0	97.5	96.2
Acid Value, mg KOH/g			
After Extraction	8.5 (17.0)	38.0 (76.0)	12.5 (25.0)

[a] Ten min in Brabender Plasticorder.

Table VII.   Carboxylation of ABS, B/AN and *cis*-1,4-Polybutadiene [a]

Polymer (35 grams)	ABS	B/AN (75/25)	Butadiene
S/MAnh (1/1 m), grams	3.6/3.4	3.6/3.4	5.4/5.1
Temperature, °C	150	150	120
Acetone Extraction			
Soluble, %	6.8	9.8	14.8
Insoluble, %	93.2	90.2	85.2
Acid Value, mg KOH/g			
After Extraction	15.5 (31.0)	28.0 (56.0)	8.0 (16.0)

[a] Ten min in Brabender Plasticorder.

Table VIII.   Carboxylation of Poly(vinyl chloride) at 180°C [a]

S/MAnh (1/1 m), grams	1.4/1.3	2.7/2.6
PVC (1% Thermolite 31), grams	40.0	40.0
Recovery, grams	41.6	44.1
Monomer Conversion, %	59.3	77.4
Acetone Extraction		
Soluble, %	35.2	34.2
Insoluble, %	64.8	65.8
Acid Value, mg KOH/g		
Before Extraction	34.4 (68.8)	54.4 (108.8)
After Extraction	17.1 (34.2)	27.7 (55.4)

[a] Two min in Brabender Plasticorder.

The carboxylation of EPR is summarized in Table VI, while the carboxylations of ABS, nitrile rubber, and *cis*-1,4-polybutadiene are shown in Table VII.

The presence of stabilizer has little effect on poly(vinyl chloride) carboxylation (Table VIII). Since the other polymers were also subjected to the carboxylation process without removing commercial additives such as antioxidants, UV and thermal stabilizers, antiblocking agents, etc.—the additives inhibit neither the polymerization of the styrene-maleic anhydride complex nor the grafting reaction.

Polystyrene, acrylic ester homopolymers and copolymers, and other polymers containing tertiary or allylic hydrogen atoms have also been readily carboxylated by the *in situ* polymerization of the styrene-maleic anhydride complex at 120 to 180°C.

Radical-catalyzed graft copolymerization is generally accompanied by crosslinking or scission (or both) of the substrate polymer. Gel permeation chromatographic analysis of polymer carboxylated by the *in situ* copolymerization of styrene and maleic anhydride without a radical catalyst shows that the molecular size distribution of the carboxylated polymer is similar to that of the original substrate polymer; this finding is consistent with the absence of radical species.

Although a radical catalyst is not necessary for graft polymerization when the alternating copolymer forms spontaneously, a radical catalyst is required when alternating copolymer formation requires radical initiation (10).

The thermal reaction of a conjugated diene and maleic anhydride yields the cyclic Diels-Alder adduct. However, alternating copolymers are readily formed when the reaction is carried out in the presence of a free-radical precursor at a temperature where the catalyst half-life is one hour or less, or the catalyst addition time is less than one hour (2, 13).

The radical-catalyzed copolymerization of isoprene and maleic anhydride in the presence of polystyrene gives the corresponding alternating copolymer graft copolymer (Table IX). Similar results are obtained with other dienes and appropriate substrate polymers.

Table IX. Graft Copolymerization of Poly(isoprene-alt-maleic anhydride) onto Polystyrene [a]

I-MAnh/PS wt ratio	1.7	4.1	8.3
Polystyrene (PS), grams	10.0	10.0	2.0
M_v	10,800	10,800	40,500
I/MAnh (1/1 m), grams	0.68/0.98	1.7/2.45	6.8/9.8
tert-BPP (75%), grams	0.5	1.2	4.8
Monomer Conversion, %	54	89	90
Fractionation (Benzene)			
P(I-MAnh), %	0	16	71
Graft Copolymer, %	100	84	29
PS/I-MAnh, mole %	89/11	87/13	28/72

[a] tert-BPP in I added over 15 min to PS + MAnh in 25 ml dioxane at 65–75°C; total reaction time, two hr

Grafting in the Presence of Metal Halide Complexing Agents

Alternating copolymer graft copolymers are also produced when comonomers that yield alternating copolymers with but not without a complexing agent are copolymerized in the presence of a complexing agent and a suitable polymer, either with or without a free-radical precursor (14).

Grafting occurs when the alternating copolymer is prepared under conditions normally used in the absence of a substrate polymer. Presence of a radical catalyst is therefore not necessary when the reaction is carried out in bulk or in an organic medium, so long as concentration of the complex is high enough. The concentration of complexing agent is 5 to 50 mole %, based on the concentration of the acceptor monomer.

Solvent concentration is kept at a level that ensures efficient stirring but low enough to avoid dilution; the temperature is generally kept below 60°C.

Table X shows the results obtained when polystyrene is dissolved in styrene monomer, followed by the addition of methyl methacrylate and EASC. Depending upon monomer concentration and after the indicated reaction period, the reaction mixture is converted to a highly swollen gel or solid product. The alternating and equimolar composition of the products was confirmed by NMR and elementary analyses, respectively.

Table X. Graft Copolymerization of Poly(styrene-alt-methyl methacrylate) onto Polystyrene in Bulk

PS, grams	3.0	3.0	3.0	3.0
S/MMA, mmoles	30/30	90/30	30/90	30/30
EASC, mmoles	15	15	45	15
Temp/Time, °C/hr	20/3	20/3	20/3	60/1.67
Monomer Conversion %	81.9	88.2	95.3	55.7
Fractionation (Cyclohexane)				
Soluble, %	32	32	29	38
PS/S-MMA, weight %	93/7	87/13	94/6	95/5
Insoluble, %	68	68	71	62
PS/S-MMA, weight %	13/87	12/88	10/90	14/86

Table XI. Graft Copolymerization of Poly(styrene-alt-acrylonitrile) onto Poly(butyl acrylate)

PBA, grams	8.0 (Solid)	8.0 (Latex)
S/AN, mmoles	125/125	125/125
$ZnCl_2$, mmoles	6.25	31.25
$AN/ZnCl_2$ mole ratio	20/1	4/1
Medium, ml	Toluene, 60	water, 10.8
Catalyst	*tert*-BPP, 0.5 ml	KPS, 0.4 g
Temp/Time, °C/hr	30/20	25/20
Monomer Conversion, %	40.8	51.0
Fractionation (Acetone)		
Soluble, %	18	25
PBA/S-AN, weight %	4/96	0/100
Insoluble, %	82	75
PBA/S-AN, weight %	44/56	48/52

The use of a radical catalyst increases polymerization rate and the extent of grafting. When the reaction is done in an aqueous medium, as when the substrate polymer is charged as a latex, a water-soluble catalyst is necessary, as previously observed for the preparation of alternating copolymers in an aqueous medium (15). Although the latex coagulates when the complexing agent is added, efficient stirring of the mixture gives a smooth reaction.

The catalyzed graft copolymerization of poly(styrene-alt-acrylonitrile) onto poly(butyl acrylate) in the presence of zinc chloride, in solution, and in an aqueous medium, is summarized in Table XI.

The catalyzed graft copolymerization of styrene-methyl methacrylate-EASC and α-methylstyrene-methacrylonitrile-EASC onto nitrile rubber in solution is shown in Table XII. In addition, grafting has been done on ethylene-propylene copolymers, polybutadiene, acrylic ester copolymers, and other polymers containing labile hydrogen atoms.

Table XII. Graft Copolymerization onto Nitrile Rubber (B/AN = 70/30)

P(B/AN), grams	5.6	5.6
Monomers	S/MMA	MS/MAN
Moles	125/125	125/125
EASC, mmoles	10	25
Medium, ml	Benzene, 55	Toluene, 55
Catalyst, grams	BP, 0.25	BP, 0.25
Temp/Time, °C/hr	40/4	40/4
Monomer Conversion, %	8.6	7.5
Fractionation (Acetone)		
Soluble, %	44	40
P(B/AN), weight %	90	80
Alt, weight %	S/MMA 10	MS/MAN 20
Insoluble, %	56	60
P(B/AN), weight %	58	74
Alt, weight %	S/MMA 42	MS/MAN 26

When the polymerization is carried out below 60°C, the grafted copolymer branches have an equimolar composition. Above 60°C, and either with or without a radical catalyst, little or no grafting occurs and nonequimolar ungrafted copolymer is obtained—particularly when the polymerizing composition is styrene-acrylonitrile-EASC. Earlier work (16) has shown that this composition yields radical rather than alternating copolymer at elevated temperatures. When the polymerizing composition is styrene-acrylonitrile-zinc chloride, temperatures above 60°C give grafted copolymer rich in acrylonitrile. Acrylonitrile-rich copolymers have been obtained at elevated temperatures because of polymerization of the AN-AN-ZnCl$_2$ complex (17).

Higher levels of grafting are obtained on polymers containing polar functionality—e.g., ester or nitrile group—when such polymers are complexed before adding the comonomers and free-radical catalyst (18).

The complexing agent is added to the polymer, either in solution or dispersed or emulsified in a nonsolvent. Reaction time varies from five minutes to 48 hours, depending upon the specific functional group, its concentration and accessibility, and the reaction medium. The comonomers and the free-radical precursor are then added. In a heterogeneous system such as a polymer latex, the free-radical catalyst is usually added after the monomers have had a chance to diffuse into the precomplexed polymer.

The temperature at which the complex is formed may be the same or different from that used for the copolymerization of the monomer.

The results shown in Table XIII were obtained when solid zinc chloride was mixed with solid poly(butyl acrylate) and the mixture stirred or homogenized vigorously for one hour at 50°C (until a transparent yellow-brown gel was obtained). The gel was dissolved in 50 ml of toluene and the mixture stirred at 50°C for six hours. The styrene-acrylonitrile mixture was added at 25°C, followed by the organic peroxide catalyst. This reaction was carried out at 25°C for three hours. After precipitation with methanol, the product was fractionated by Soxhlet extraction, as summarized in Table XIII.

Table XIII. Graft Copolmerization of Poly(styrene-alt-acrylonitrile) onto Precomplexed Poly(butyl acrylate)–ZnCl$_2$

Precomplexation		*Polymerization*	
PBA, grams (mmoles)	6.5 (50)	AN, grams (mmoles)	6.6 (125)
ZnCl$_2$, grams (mmoles)	3.4 (25)	S, grams (mmoles)	13.0 (125)
Temp/Time, °C/hr	50/1		
Toluene, ml	50	MPP (50%), ml	0.5
Temp/Time, °C/hr	50/6	Temp/Time, °C/hr	25/3

Monomer Conversion 53.0%

Fractionation

Solvent	%	BA/S/AN Mole Ratio	S/AN Mole Ratio
Hexane	20.0	100/0/0	—
Ethyl Acetate	7.5	32/36/32	52/48
Residue	72.5	11/44.5/44.5	50/50

The results in Table XIV were obtained by adding an aqueous zinc chloride solution to a poly(butyl acrylate) latex. After precomplexation at 30°C, the monomer mixture and the redox catalyst were added, and the polymerization was carried out at 30°C. Because of the heterogeneity of the reaction mitxure, a large amount of alternating copolymer accompanied the alternating copolymer graft copolymer.

Table XIV.   Graft Copolymerization of Poly(styrene-alt-acrylonitrile) onto Precomplexed Poly(butyl acrylate)–ZnCl$_2$ Latex

Precomplexation		Polymerization	
PBA Latex (42.5%), grams	18.8	AN, grams (mmoles)	6.6 (125)
PBA, grams (mmoles)	8.0 (62.5)	S, grams (mmoles)	13.0 (125)
ZnCl$_2$, grams (mmoles)	1.70 (12.5)	K$_2$S$_2$O$_8$/Na$_2$S$_2$O$_3$, grams	0.27/0.19
Temp/Time, °C/hr	30/24	Temp/Time, °C/hr	30/3

Monomer Conversion 69.5%

Fractionation

Solvent	%	BA/S/AN Mole Ratio	S/AN Mole Ratio
Hexane	7.1	100/0/0	—
Acetone	47.1	2/49/49	50/50
Residue	45.8	64/18/18	50/50

Table XV.   Graft Copolymerization of Poly(styrene-alt-acrylonitrile) onto Precomplexed Poly(butyl acrylate)—ZnCl$_2$ Latex

Precomplexation		Polymerization	
PBA latex (42.5%) grams	18.8	AN, grams (mmoles)	6.6 (125)
PBA, grams (mmoles)	8.0 (62.5)	K$_2$S$_2$O$_8$, grams	0.4
Styrene, grams (mmoles)	13.0 (125)		
Temp/Time, °C/hr	30/24	Temp/Time, °C/hr	50/20
ZnCl$_2$, grams (mmoles)	4.26 (31.25)		
Temp/Time, °C/hr	42/5		

Monomer conversion 88.4%

Fractionation

Solvent	%	BA/S/AN Mole Ratio	S/AN Mole Ratio
Hexane	8.7	100/0/0	—
Acetone	7.3	1/46.5/52.5	47/53
Residue	84.0	25/40/35	53/47

The yield of graft copolymer was increased by mixing the latex with styrene for 24 hours before adding solid zinc chloride. This permitted easier diffusion of the latter in the polymer. As a result (*see* Table XV), monomer conversion was increased, yield of ungrafted alternating copolymer was reduced, and yield of graft copolymer was increased.

Precomplexation is not as effective in a homogeneous system, where the complexing agent has an opportunity to diffuse away from the polymer and complex with the acceptor monomer. This is particularly true when precomplexation time is short (Table XVI).

Table XVI. Graft Copolymerization of Poly(styrene-alt-acrylonitrile) onto Precomplexed Poly(butyl acrylate)—EASC

Precomplexation		*Polymerization*	
PBA, grams (mmoles)	8.0 (62.5)	AN, grams (mmoles)	6.6 (125)
EASC, ml (mmoles)	3.5 (15.7)	S, grams (mmoles)	13.0 (125)
Toluene, ml	50	MPP (50%), ml	0.5
Temp/Time, °C/hr	25/0.5	Temp/Time, °C/hr	22/3

Monomer conversion 55.6%

	Fractionation		
Solvent	*%*	*BA/S/AN Mole Ratio*	*S/AN Mole Ratio*
Hexane	3.6	100/0/0	—
Ethyl acetate	64.9	2/52.5/45.5	54/46
Residue	31.5	67/18/15	54/46

Table XVII shows the grafting of poly(styrene-alt-acrylonitrile) on nitrile rubber precomplexed with EASC. This system is heterogeneous since addition of EASC to the polymer solution causes the polymer to agglomerate or precipitate.

An effective method of preparing precomplexed polymer is to homopolymerize a polar monomer in the presence of a complexing agent using light, a small amount of oxygen, or a free-radical catalyst for initiation. The homopolymer contains complexed complexing agent. When a mixture of comonomers is added to the complexed polymer, followed by addition of a free-radical catalyst (if necessary), alternating copolymer and graft copolymer are obtained.

Similarly, an alternating copolymer containing complexed complexing agent may be prepared from a comonomer mixture and then mixed with different comonomers to form an alternating copolymer graft copolymer.

Table XVII. Graft Copolymerization of Poly(styrene-alt-acrylonitrile) onto Precomplexed Nitrile Rubber–EASC

Precomplexation		Polymerization	
P(B/AN) (70:30), grams	4.0	AN, grams (mmoles)	4.1 (77)
EASC, mmoles	22	S, grams (mmoles)	8.0 (77)
Toluene, ml	40	Benzoyl Peroxide, grams	0.36
Temp/Time, °C/hr	40/1	Temp/Time, °C/hr	40/5

Monomer conversion 87%

	Fractionation		
Solvent	%	B/AN/S Mole Ratio	S/AN Mole Ratio
Benzene	19	64/27/9	9/0
Acetone	12	0/50/50	50/50
Residue	69	40/39/21	49/51

Grafting on Cellulose and Polyvinyl Alcohol

Formation of alternating copolymer graft copolymers from binary monomer mixtures suggests that cellulose acts as a matrix and promotes the formation of comonomer charge transfer complexes (19). This is in spite of the fact that graft copolymerization of acrylic monomers on cellulose in aqueous suspension, either without a catalyst, under irradiation, or in the presence of oxidizing agents is considered to involve a radical mechanism complicated by diffusion effects stemming from the reaction mixture's heterogeneity and high viscosity.

The copolymerization of styrene and methyl methacrylate (90/10 molar ratio) in the presence of wood pulp at 90°C using $NaClO_2$ as catalyst gives equimolar grafted and ungrafted copolymers at a low conversion. However, while the grafted copolymer composition is still equimolar at a higher conversion, the ungrafted copolymer composition approaches that expected from the comonomer charge in accordance with a radical mechanism (20).

When a mixture of butadiene and methyl methacrylate is heated at 90°C with an aqueous suspension of kraft wood pulp containing a small amount of nonionic surfactant and in the absence of a catalyst, essentially equimolar alternating butadiene-methyl methacrylate copolymer is grafted on the cellulose when the monomer charge contains 20–50 mole % methyl methacrylate (21).

The mutual γ-irradiation of viscose rayon immersed in a methanol solution containing butadiene and acrylonitrile at 30°C yields ungrafted copolymers having the compositions expected for a normal radical co-polymerization over a wide range of monomer charge ratios. However, the grafted copolymers are essentially equimolar when the acrylonitrile content in the monomer charge is from 30 to 90 mole % (*22*).

Under similar conditions, graft polymerization of a mixture of sty-rene and acrylonitrile on viscose rayon yields equimolar grafted copoly-mers and ungrafted copolymers with a radical composition when the monomer charge contains 25 to 70 mole % acrylonitrile.

Similar results are obtained in the radiation-induced graft copoly-merization of butadiene-acrylonitrile and styrene-acrylonitrile onto poly-vinyl alcohol fibers (*22*).

Cellulose-water may act as a matrix and promote the development of arrays of comonomer charge transfer complexes (*19*). The cellulose acts not only as a substrate for such alignment but also as a complexing agent. The matrix of complexes may be represented as shown in I (styrene-methyl methacrylate) and II (butadiene-acrylonitrile). The radical-, thermal-, and radiation-induced graft polymerizations involve homopolymerization of comonomer complexes rather than copolymeriza-tion of uncomplexed monomers.

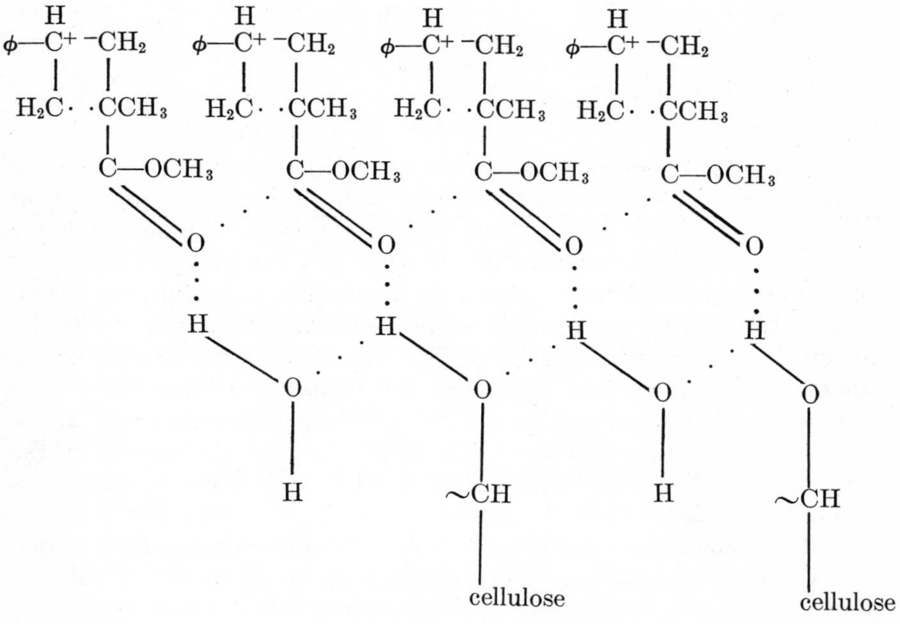

I

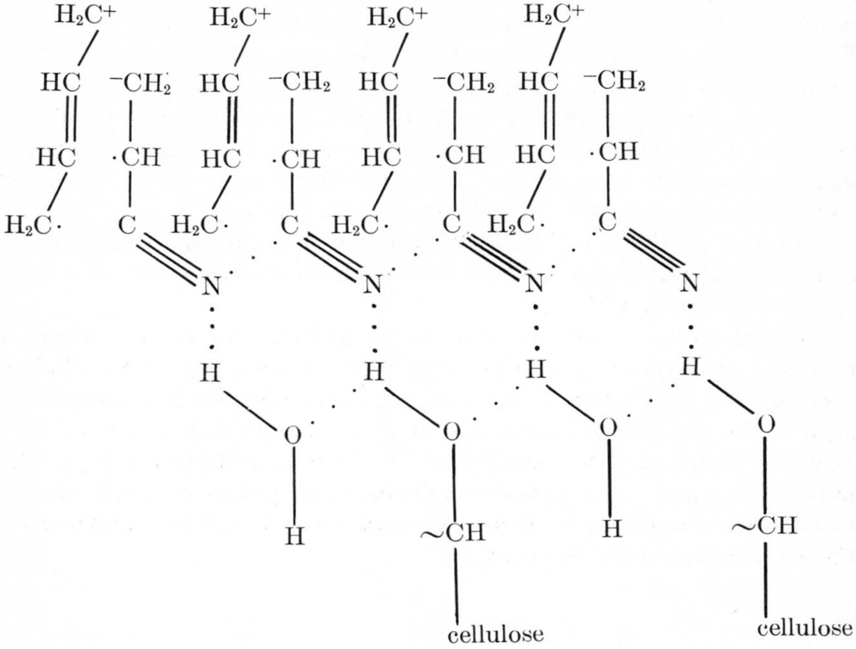

H₂C⁺ .... H₂C⁺ .... H₂C⁺ .... H₂C⁺

HC   −CH₂   HC   −CH₂   HC   −CH₂   HC   −CH₂

HC   ·CH   HC   ·CH   HC   ·CH   HC   ·CH

H₂C·   C   H₂C·   C   H₂C·   C   H₂C·   C

N   N   N   N

H   H   H   H

O   O   O   O

H   ~CH   H   ~CH

cellulose   cellulose

II

When the copolymerization of styrene and acrylonitrile in the presence of zinc chloride is carried out at 40°-50°C in the presence of cellulose under conditions that normally yield an alternating copolymer (*i.e.*, in an aqueous system in the presence of potassium persulfate or a persulfate-bisulfite redox system), the alternating copolymer is accompanied by cellulose graft copolymer in which the grafted chains have an equimolar, alternating structure. The monomer conversion is higher than when the reaction is carried out without cellulose, and both ungrafted and grafted copolymers (60-80% of the total copolymer) are of very high molecular weight (Table XVIII). When the reaction in the presence of cellulose is conducted above 60°C, normal radical copolymer is produced (*23*).

The comonomer complexes may be anchored on the cellulose, analogous to structures I and II, or through the interaction of zinc chloride and the cellulosic hydroxyl groups; aqueous solutions of the metal halide are known to break the hydrogen bonds in cellulose and reduce crystallinity. Although radicals generated on the cellulose as the result of reaction with the catalyst may initiate polymerization of the comonomer

Table XVIII.   Copolymerization of Styrene and Acrylonitrile in Presence of ZnCl₂ and Cellulose

Temperature, °C	40	40	50
Time, hr	3	5	1
Monomer Conversion, %	22	32	16
Add-on, %	221	353	146
Grafted copolymer			
% of Total	63.9	66.7	58.6
AN, mole %	50.0	49.0	50.2
[η] dl/g	5.4	5.3	5.2
Ungrafted Copolymer			
% of Total	36.1	33.3	41.4
AN, mole %	50.0	50.2	52.0
[η] dl/g	5.1	4.3	3.2

[S] = [AN] = 1.0 mole; [ZnCl₂] = 0.5 mole; [K₂S₂O₈] = 0.05 mole; H₂O, 157 grams; wood pulp, 10 grams; [η] DMF, 30°C.

charge-transfer complexes, they are not the sites for the attachment of grafted alternating copolymer chains.

There is a maximum of one grafted chain per cellulose or polyvinyl alcohol molecule, independent of the initiation method (*24*). A relationship between the aldehyde content and the extent of grafting has also been demonstrated (*19*).

It is therefore reasonable to assume that grafting occurs by termination of a propagating alternating copolymer chain on the substrate

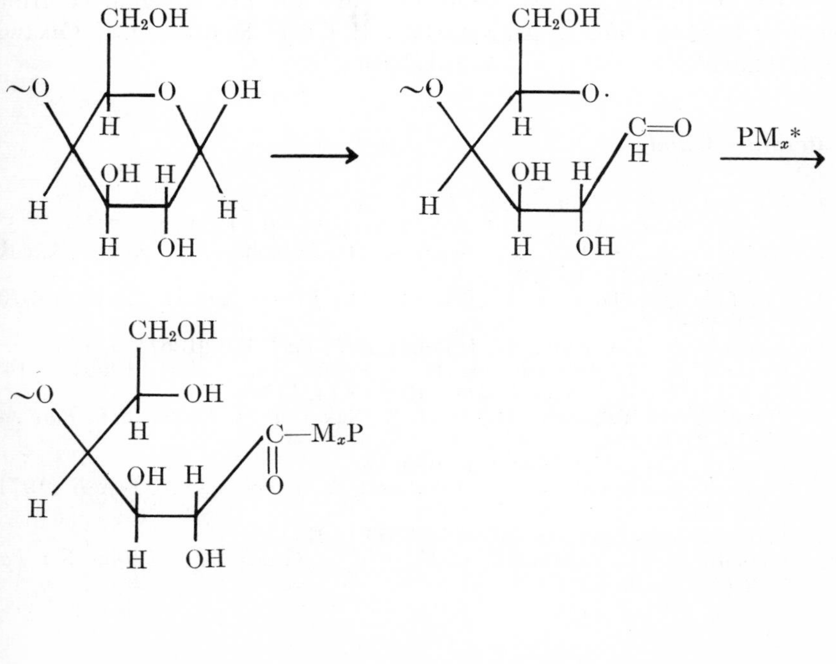

$$\underset{OH}{\sim CH_2CHCH_2CH} - \underset{OH}{CHCH_2CH_2CH} \sim \; \longrightarrow \; \underset{OH}{\sim CH_2CHCH_2CH} + \cdot \underset{OH}{\overset{H}{CCH_2CH_2CH}} \sim$$

$$PM_x{}^*\downarrow$$

$$\underset{OH}{\sim CH_2CHCH_2C} - M_xP$$

polymer, presumably by insertion of the generated terminal carbene into the C–H of an aldehyde group. In cellulose, the latter is apparently a terminal group generated by cleavage of the terminal hemiacetal groups. In polyvinyl alcohol, the aldehyde is a terminal group generated by cleavage of the vicinal glycol resulting from head-to-head addition. Although aldehyde groups are also generated by cleavage of the vicinal glycol in the anhydroglucose units of cellulose, these groups are relatively inaccessible and are buried in the crystalline hydrogen-bonded cellulose, compared with the terminal aldehydes which are probably in disordered areas. The alternating copolymer graft copolymers on cellulose and polyvinyl alcohol are thus actually alternating copolymer block copolymers.

Acknowledgment

The author is pleased to acknowledge the experimental contributions of L. C. Anand, H. Antropiusova, R. Guzzi, S. Kikuchi, E. Oikawa, B. K. Patnaik, M. Stolka, and A. Takahashi.

Literature Cited

1. Gaylord, N. G., J. Polym. Sci., Part C (1970) 31, 247.
2. Gaylord, N. G., Maiti, S., J. Polym. Sci., Part B (1971) 9, 359.
3. Tsuchida, E., Ohtani, Y., Nakadai, H., Shinohara, I., Kogyo Kagaku Zasshi (1967) 70, 573.
4. Tsuchida, E., Tomono, T., Nishide, H., Kogyo Kagaku Zasshi (1970) 73, 2037.
5. Iwatsuki, S., Yamashita, Y., J. Polym. Sci., Part A-1 (1967) 5, 1753.
6. Gaylord, N. G., Antropiusova, H., J. Polym. Sci., Part B (1969) 7, 145
7. Frank, H. P., Makromol. Chem. (1968) 114, 113.
8. Hirooka, M., Yabuuchi, H., Iseki, J., Nakai, Y., J. Polym. Sci., Part A-1 (1968) 6, 1381.
9. Gaylord, N. G., U.S. patent pending.
10. Gaylord, N. G., Oikawa, E., Takahashi, A., J. Polym. Sci., Part B (1971) 9, 379.
11. Gaylord, N. G., Belgian Patent 762,854 (1971).
12. Gaylord, N. G., Takahashi, A., Kikuchi, S., Guzzi, R., J. Polym. Sci. Part B (1972) 10, 95.

13. Gaylord, N. G., Stolka, M., Takahashi, A., Maiti, S., *J. Macromol. Sci. (Chem.)* (1971) **A5**, 867.
14. Gaylord, N. G., Antropiusova, H., Patnaik, B. K., *J. Polym. Sci., Part B* (1971) **9**, 387.
15. Gaylord, N. G., Takahashi, A., Anand, L. C., *J. Polym. Sci. Part B* (1971) **9**, 97.
16. Gaylord, N. G., Patnaik, B. K., *J. Polym. Sci., Part B* (1970) **8**, 411.
17. Gaylord, N. G., Antropiusova, H., *J. Polym. Sci., Part B* (1970) **8**, 183.
18. Gaylord, N. G., Patnaik, B. K., Stolka, M., *J. Polym. Sci., Part B* (1971) **9**, 923.
19. Gaylord, N. G., *J. Polym. Sci., Part C* (1972) **37**, 153.
20. Rebek, M., Schurz, J., Stöger, W., Popp, W., *Monatsh. Chem.* (1969) **100**, 532.
21. Nara, S., Matsuyama, K., *Kobunshi Kagaku* (1968) **25**, 840.
22. Sakurada, I., Okada, T., Hatakeyama, S., Kimura, F., *J. Polym. Sci., Part C* (1963) **4**, 1233.
23. Gaylord, N. G., Anand, L. C., *J. Polym. Sci., Part B* (1971) **9**, 617.
24. Sakurada, I., Ikada, Y., Uehara, H., Nishizaki, Y., Horii, F., *Makromol. Chem.* (1970) **139**, 183.

RECEIVED April 1, 1972.

14

New Block Copolymers from Macroradicals

RAYMOND B. SEYMOUR, PATRICK D. KINCAID,
and DONALD R. OWEN

University of Houston, Houston, Tex. 77004

New macroradicals have been obtained by proper solvent selection for the homopolymerization of styrene, methyl methacrylate, ethyl acrylate, acrylonitrile, and vinyl acetate, and by the copolymerization of maleic anhydride with vinyl acetate, vinyl isobutyl ether, or methyl methacrylate. These macroradicals and those prepared by the addition to them of other monomers were stable provided they were insoluble in the solvent. Since it does not add to maleic anhydride chain ends, acrylonitrile formed a block copolymer with only half of the styrene-maleic anhydride macroradicals. However, this monomer gave excellent yields of block polymer when it was added to a macroradical obtained by the addition of limited quantities of styrene to the original macroradical. Because of poor diffusion, styrene did not add to acrylonitrile macroradicals, but block copolymers formed when an equimolar mixture of styrene and maleic anhydride was added.

It has previously been shown that the rate of the free-radical-initiated copolymerization of maleic anhydride and vinyl monomers is faster in poor solvents than in good solvents, and that the copolymerization rates in these poor solvents decrease as the difference in the solubility parameter values between that of the copolymer and the solvent increases (*6, 10*). Other investigators have suggested that this free-radical initiation involves hydrogen abstraction, and that the free-radical precursor increases the yield of the alternating copolymer regardless of the monomer charge (*3*).

The tendency for these monomers to produce an alternating copolymer is also supported by their reactivity ratios—0 for maleic anhydride and 0.02 for styrene. It has also been suggested that a strong donor monomer such as maleic anhydride and a strong acceptor monomer such as styrene produce a comonomer charge-transfer complex which, in its excited state, is the polymerizable specie (*4, 5*).

While it is assumed that termination by coupling takes place when maleic anhydride and styrene are copolymerized in a good solvent such as acetone, insoluble macroradicals precipitate when these monomers are copolymerized in a poor solvent such as benzene (*7*). Insoluble macroradicals obtained by bulk polymerization of acrylonitrile (*1, 11*) and the solution copolymerization of maleic anhydride and styrene in benzene (*7*) have been used as seeds for the preparation of block copolymers.

Also, it has been reported that poly(styrene-co-maleic anhydride-b-styrene) could be obtained either by addition of styrene monomer to the styrene-maleic anhydride macroradicals or by the free-radical-initiated copolymerization of maleic anhydride with more than an equimolar proportion of styrene in benzene solution (*7*). However, the maximum amount of styrene present in these block copolymers was less than 35% of the weight of the original macroradical. This limitation on the yield of the block copolymer is now assumed to be related to the increased solubility of the styrene block in the benzene solvent, which permits termination of the new macroradicals by coupling.

Block Copolymers from Styrene-Maleic Anhydride Copolymer Macroradicals

No product was obtained when equimolar mixtures of maleic anhydride and styrene in a benzene solution were heated at 50°C either without oxygen or with large amounts of oxygen. However, a benzene-insoluble product was obtained when air was not removed from these solutions, and the sealed containers in the presence of a moderate amount of oxygen were heated at 50°C.

Yields in excess of 98% of macroradicals were obtained in less than two hours when equimolar oxygen-free mixtures of maleic anhydride and styrene were heated at 50°C in the presence of 2.5% α, α'-azobisisobutyronitrile (AIBN). These macroradicals could be isolated by filtering in an inert atmosphere and removing residual solvent by applying a vacuum at 0°C. The macroradicals retained their ability to form block copolymers after being stored for seven days in the absence of oxygen at 0°C.

Figure 1 shows that the weight of these macroradicals increases by about 35% when they are heated for 50 hours with relatively large amounts of styrene in a benzene solution at 50°C. However, a weight increase of more than 300% was noted when the macroradicals were heated with an equimolar mixture of maleic anhydride and styrene in a benzene solution at 50°C. The presence of macroradicals was confirmed by the formation of high yields of block copolymers when acrylonitrile was added to the macroradicals at 50°C (9). However, no increase in weight was noted when either a mixture of maleic anhydride and styrene or acrylonitrile was added to the block copolymer obtained by the addition of at least 35% styrene.

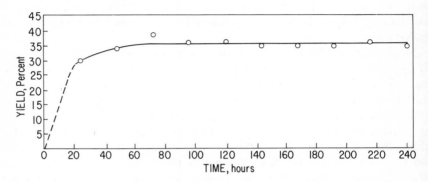

Figure 1. Rate of addition of styrene monomer to a styrene-maleic anhydride macroradical in benzene at 50°C

Block copolymers in good yield were obtained when the styrene-maleic anhydride macroradicals were heated with methyl methacrylate in benzene at 50°C. (*see* Figure 2).   Unlike what happened when styrene monomer was added, the yield of the poly(styrene-co-maleic anhydride-b-methyl methacrylate) continued to increase with time up to 180 hours.   The relative areas of the styrene and methyl methacrylate pyrolysis gas-chromatographic (PGC) peaks were 0.53 and 0.24 for the block copolymers containing 17 and 35% methyl methacrylate, respectively.

The rate of block copolymer formation in a poor solvent is inversely related to the difference between the solubility parameters of the macroradical and the monomer used to form the block.   Thus, when these solubility parameter values are similar—as with the addition of styrene and maleic anhydride to the styrene-maleic anhydride macroradical—a 100% weight increase is observed in a few hours.

Solubility parameters are a measure of polarity that give a quantitative measurement to "like dissolves like." They vary from 6 for nonpolar aliphatic hydrocarbons to 23.2 for water.   This well-known parameter is adequately described in other publications (2, 8).

When the difference in the solubility parameters is greater than 2 hildebrand units, as in the case with methyl methacrylate and the styrene-maleic anhydride macroradical, the rate of formation of block

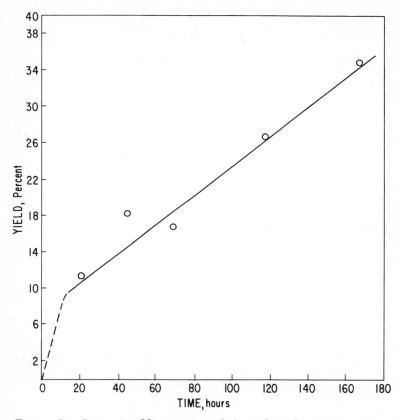

Figure 2.   Rate of addition of methyl methacrylate monomer to styrene-maleic anhydride macroradicals in benzene at 50°C

copolymers is slower. Thus, as shown in Figure 2, only 11 parts by weight of this monomer are added to the 100 parts of the macroradical during 24 hours.

The difference between the solubility parameters of styrene and these macroradicals is 1.7 hildebrand units. The weight of the macroradical thus increased by 30 percent in 24 hours in the presence of styrene (Figure 1).

The difference between the solubility parameter of acrylonitrile and the styrene-maleic anhydride macroradical is 0.5 hildebrand unit. The formation of block copolymer was thus rapid, and the weight of these macroradicals increased by 86 percent in 24 hours in the presence of acrylonitrile (Figure 3).

An increase in weight was noted when an equimolar mixture of styrene and maleic anhydride was added to these acrylonitrile block macroradicals.   The relative areas of the styrene/acrylonitrile peaks of the macroradical and the copolymer with the styrene-co-maleic anhydride block were 5.82 and 6.45, respectively.

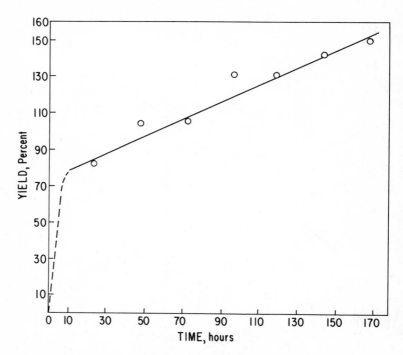

Figure 3.   Rate of addition of acrylonitrile monomer to a styrene-maleic anhydride macroradical in benzene at 50°C

Acrylonitrile was also added to a mixture of benzene and macroradicals obtained from a monomer mixture containing equimolar amounts of maleic anhydride and styrene.   Compared with the data already cited for styrene-rich macroradicals, only 50% of the macroradicals obtained from the equimolar mixture produced acetone-insoluble block copolymers. Since a mixture of maleic anhydride and acrylonitrile did not form a copolymer when heated with AIBN in benzene, it was concluded that half of these original macroradicals had maleic anhydride terminal groups.

Dead copolymers, as noted, were obtained when large amounts of styrene monomer were added to styrene-maleic anhydride macroradicals. However, macroradicals were obtained when the amount of styrene added equalled less than 30% of the weight of the macroradical.   For example, block copolymers were obtained when styrene and maleic anhydride or acrylonitrile were added to styrene-co-maleic anhydride-b-styrene macroradicals in which the ratio of the weight of the styrene block to that of the original macroradical was 27/100.   Poly(styrene-co-maleic anhydride-b-styrene-b-vinyl acetate) was also obtained.

Block Copolymers from Other Macroradicals Obtained from Maleic Anhydride and Vinyl Monomers

Macroradicals were also prepared by the copolymerization of maleic anhydride and vinyl acetate in benzene in the presence of 2.5% AIBN. Because of the solubility of the vinyl acetate block in benzene, the maximum ratio of the weight of the vinyl acetate to the macroradical in poly(vinyl acetate-co-maleic anhydride-b-vinyl acetate) was 26/100.   By contrast, since the acrylonitrile block was insoluble in benzene, excellent yields of poly(vinyl acetate-co-maleic anhydride-b-acrylonitrile) were obtained.   For example, the ratio of the weight of the acrylonitrile to that of the macroradical after 10 days in benzene at 50°C was 131/100. Macroradicals were also prepared by the copolymerization of maleic anhydride and vinyl isobutyl ether in benzene with 2.5% AIBN.

Since the reactivity ratios of maleic anhydride and methyl methacrylate are 0.03 and 3.5, respectively, there is less tendency for the forma-

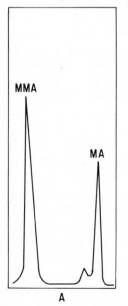
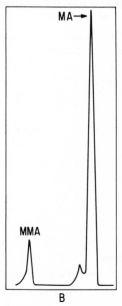

Figure 4.   Chromatograms of pyrolyzates from copolymers of methyl methacrylate (MMA) with equimolar concentration of maleic anhydride (MA) [A] and with fourfold molar excess of MA [B]

tion of an alternating copolymer than in the previous case with styrene. Thus, when equimolar ratios of maleic anhydride and methyl methacrylate were copolymerized in benzene in the presence of 2.5% AIBN at 50°C, a dead-soluble random copolymer containing much less than 50% maleic anhydride was obtained.

However, an active benzene-insoluble methyl methacrylate-maleic anhydride macroradical was obtained when the molar ratio of the maleic anhydride to methyl methacrylate was 4 to 1. This random macroradical contained about 50% maleic anhydride. Excellent yields of block copolymers were obtained when mixtures of maleic anhydride and styrene or vinyl acetate were added to these macroradicals. For example, the ratios of the weight of the styrene-maleic anhydride and the vinyl acetate-maleic anhydride blocks to the weight of the macroradical were 790/100 and 627/100, respectively, after the mixtures of the monomers and the macroradicals were heated in benzene for three days at 50°C.

The relative areas of the GC chromatograms obtained from the pyrolyzates from the dead methyl methacrylate-maleic anhydride copolymer and the corresponding macroradical are shown in Figure 4.

Preparation of Block Copolymers from Homomacroradicals

Macroradicals were obtained by the polymerization of ethyl acrylate in cyclohexane, styrene in hexane, vinyl acetate in decane, and methyl methacrylate in hexane. Because of the solubility of the vinyl acetate block in hexane, the ratio of the weight of vinyl acetate to that of the macroradical in poly(methyl methacrylate-b-vinyl acetate) after heating at 50°C for three days was only 30/100. By contrast, because of the insolubility of the acrylonitrile block in hexane, good yields of methyl methacrylate-b-acrylonitrile macroradicals were obtained. The ratio of the weight of the acrylonitrile block to that of the macroradical was thus 90/100 after heating the mixture for three days at 50°C in hexane.

Good yields of acrylonitrile macroradicals were obtained when a benzene solution of this monomer was heated at 50°C with 2.5% AIBN. The difference between the solubility parameters of these macroradicals and styrene, 3.2 hildebrand units, was too great to permit diffusion of the styrene monomer in the acrylonitrile macroradical and therefore, no block copolymer was produced with this monomer.

However, the difference between the solubility parameters of an equimolar mixture of styrene and maleic anhydride and the acrylonitrile macroradical is only 1.1 hildebrand units. Thus, a block copolymer with a ratio by weight of styrene and maleic anhydride to that of the acrylonitrile macroradical of 41/100 was obtained when a mixture of these monomers and the macroradicals were heated in benzene at 50°C for 24 hours.

The formation of block copolymers was demonstrated by an increase in weight that did not take place either with homogeneous solution polymers or with heterogeneous solution polymers in the presence of oxygen. That these high-yield products were not mixtures of homopolymers was obvious from their solubility characteristics. However, this conclusion was also verified by solvent extraction and pyrolysis of the solvent fractions. The presence of more than one monomer in the pyrolyzate was demonstrated by characteristic gas-chromatographic retention times.

For example, the styrene-maleic anhydride copolymer is soluble in acetone, and polyacrylonitrile is insoluble in acetone but soluble in dimethylformamide. Yet, the block copolymer obtained by addition of acrylonitrile to the styrene-maleic anhydride macroradical is soluble neither in acetone nor in dimethylformamide. Pyrolysis gas-chromatography, though, shows that this acetone and dimethylformamide-insoluble product contains styrene, maleic anhydride, and acrylonitrile. The relative area of the acrylonitrile peak was related to the amount of acrylonitrile added to the original macroradical.

Experimental

Maleic anhydride was crystallized from sodium-dried, thiophene-free benzene. The azobisisobutyronitrile was crystallized from a mixed solvent containing equal volumes of benzene and toluene. The liquid monomers were purified by distillation under reduced pressure.

To prepare the copolymers, enough solvent was added to 0.5 gram of equimolar quantities of maleic anhydride and vinyl monomer and 0.0125 gram azobisisobutyronitrile to make a total volume of 5 ml. The copolymerizations were conducted in sealed glass containers in the absence of oxygen. The precipitate was removed from heterogeneous systems, washed well with fresh solvent after filtration, and dried under a vacuum at 0°C.

The homopolymers were prepared similarly, using solvents in which the polymers were insoluble. The block copolymers were produced by adding appropriate monomers to the mixture of the macroradicals and benzene, and heating the mixture in an inert atmosphere at 50°C.

Pyrolysis gas-chromatographic investigations were made using a Wilkins model A100C Aerograph equipped with a Servo-Ritter II Texas Instruments recorder; helium was the carrier gas. The 10-foot by 1/4-inch diameter column was packed with acid-washed Chromosorb W (Johns Manville) with 20% SE-20 (General Electric).

A rhenium tungsten pyrolyzing coil (code 13-002) obtained from Gow-Mac Instrument was used. Solutions of the samples were placed on the coils and the solvent allowed to evaporate before the residual film was pyrolyzed for about five seconds.

Conclusions

Macroradicals can be prepared by free-radical-initiated solution polymerization of monomers in poor solvents. Monomers with solubility parameters similar to those of the macroradicals may form block copolymers in solvents that are poor solvents for both the macroradical and the block. The ability of a block macroradical to add an additional block is governed by the solubility parameter of the initial chain in the macroradical, and not by the solubility parameter of the end block. For the formation of macroradicals, it is essential that the solubility parameters of the monomer and polymer differ by at least 1.8 hildebrand units. For the formation of block copolymers, it is essential that the difference in solubility parameters of the monomer and macroradical be less than 3.2 hildebrand units.

Literature Cited

1. Bamford, C. H., Jenkins, A. D., *J. Chim. Phys.* (1959) **56**, 798.
2. Burrell, H., "Polymer Handbook," Interscience, New York (1963).
3. Gaylord, N. G., *J. Polym. Sci.* (1970) **3K**, 247.
4. Gaylord, N. G., Stalka, M., Takahashi, A., Maiti, S., *J. Macromol. Sci. (Chem)* (1971) **A5**, 867.
5. Gaylord, N. G., Maiti, S., *J. Polym. Sci.* (1971) **9B,** 359.
6. Seymour, R. B., Tatum, S. D., Boriack, C. J., Tsang, H. S., *The Texas J. Sci.* (1969) **21(1)** 13.
7. Seymour, R. B., Tsang, H. S., Jones, E. E., Kincaid, P. D., Patel, A. K., ADVAN. CHEM. SER. (1971) **99,** 418.
8. Seymour, R. B., *Modern Plastics* (1971) **48(10)** 150.
9. Seymour, R. B., Kincaid, P. D., *J. Paint Technol.* (1973) **45** (580) 33.
10. Tsuchida, E., Ohtani, U., Nakadai, H., Shinohara, I., *Chem. Soc. Japan J.* (1967) **70**, 573.
11. Yugi, M., Yayoi, O., *J. Polym. Sci.* (1969) **7A-1,** 2547.

RECEIVED April 13, 1972.

15

Polystyrene–Polydimethylsiloxane Multiblock Copolymers

JOHN C. SAAM, ANDREW H. WARD and F. W. GORDON FEARON

Dow Corning Corp., Midland, Mich., 48640

The title block copolymers with characteristics ranging from thermoplastic elastomers to polyethylene-like thermoplastics are obtained from ring opening polymerization of hexamethylcyclotrisiloxane with "living" α,ω-dilithiopolystyrene. Chain scissions and oligomerizations which usually complicate siloxane polymerization are avoided, and molecular parameters regulating physical and mechanical properties are conveniently controlled to provide a unique family of thermoplastic materials.

Block copolymers of polydimethylsiloxane and various thermoplastics offer combinations of properties which have intrigued numerous investigators. The interest generated by the earlier copolymer systems as well as those of the present investigation stems in large part from the unique features imparted by the polydimethylsiloxane blocks. Among these are retention of flexibility at low temperature, excellent electrical properties, ozone resistance, durability towards weathering, and a high degree of permeability towards gases. Thermoplastic blocks reported to be incorporated in such thermoplastic elastomers so far include silphenylene–siloxane (*1*), poly(bisphenol-A–carbonate) (*2*), polystyrene (*3, 4*), and polyarylsulfones (*5*). In each case the rubbery part of the block copolymer is polydimethylsiloxane. All show interesting and unique property profiles, but the only block copolymers which seem economically suited for volume manufacture are those where the "hard" blocks are composed of polystyrene or its derivatives.

The most promising approach for preparing block copolymers of polystyrene (A) and polydimethylsiloxane (B) involves polymerization of cyclosiloxane monomers with "living" α,ω-polystyrene anions (*6*). The original approach, however, gives materials contaminated with the component homopolymers which are ruinous to mechanical properties. The present effort shows that variation of the original anionic polymerization of cyclosiloxanes with living polystyrene (7) circumvents these difficulties and provides an effective route to a range of useful materials.

Synthesis

The key to synthesizing well defined block copolymers successfully is the anionic ring-opening polymerization of hexamethylcyclotrisiloxane, D_3, from the "living" ends of α,ω-dilithiopolystyrene. This gives a BAB block copolymer terminated with lithium silanolate ends. The ends are then coupled with dialkyldichlorosilane to give $(BAB)_x$ multiblock copolymers essentially free of homopolymers and by-product cyclosiloxanes. The siloxane polymerization is run in solution in the presence of promoting solvents such as THF or the glymes. Contrary to typical anionic cyclosiloxane polymerizations where potassium silanolates are usually employed, little or no concomitant chain scission or equilibration is observed when lithium is the counterion. This has been demonstrated in model experiments where butyllithium rather than "living" polystyrene was used to polymerize D_3 under comparable conditions. These gave little of the by-product cyclodimethylsiloxanes usually found in anionic siloxane ring-opening polymerizations done in solution. High conversions to polydimethylsiloxane with narrow molecular weight distribution were also obtained. The molecular weight corresponded closely to that calculated from the amount of catalyst and the weight of D_3 used provided the system was scrupulously purged of moisture and other protic impurities. Thus, as in "living" styrene polymerizations the molecular weight of the siloxane blocks can be closely regulated. This procedure allows control of block size and relative amount of the blocks in the copolymer system. The precise nature of the polymerization and the ability to control molecular variables were then used to synthesize polystyrene–polydimethylsiloxane AB block copolymers (8). Reaction 1 outlines the synthesis of the multiblock copolymers.

$$\text{Li[CH}_2\text{CH(C}_6\text{H}_5\text{)]}_n\text{Li} + 2/3\text{M(Me}_2\text{SiO)}_3 + \text{(polar solvent)} \rightarrow$$

$$\text{Li(OMe}_2\text{Si)}_m\text{[CH}_2\text{CH(C}_6\text{H}_5\text{)]}_n\text{(SiMe}_2\text{O)}_m\text{Li} \qquad (1)$$

$$\Big\downarrow \text{R}_2\text{SiX}_2$$

$$\text{[(OMe}_2\text{Si)}_m\text{(CH}_2\text{—CH(C}_6\text{H}_5\text{))}_n\text{(SiMe}_2\text{O)}_m]_x$$

The experimental conditions were essentially those already reported for the synthesis of the corresponding AB block copolymers. The only differences were the difunctional initiator (9) and the use of a difunctional rather than monofunctional chlorosilane for reaction with the lithium siloxanolate terminated polymer. At least 18 hours were allowed for the latter step to ensure complete reaction. Absence of homopolymer was assessed by the previously described procedure of determining solubility in selective solvents (8) and in two examples by fractional precipitation of 1% toluene solutions of block copolymer using methanol as a precipitant. The compositions of each fraction as determined by silicon analysis were constant within experimental error over the range of molecular weights obtained (given in Figures 1 and 2.) Gross variations in silicon content from one fraction to another would be expected for either the copolymer rich in polydimethylsiloxane or the copolymer rich in polystyrene if significant amounts of either homopolymer were present.

Table I. Effect of Free Polydimethylsiloxane[a] on Mechanical Properties of a Compression Molded (BAB)$_x$ Block Copolymer with 50 wt % Polystyrene[b]

$(Me_2SiO)_n$ Added, wt %	Tensile at Break, psi, (Break)	Elongation at Break, %, (Break)
0	2200	350
5	1600	280
10	1400	340
15	440	40
20	150	20

[a] Molecular weight 22,800.
[b] Molecular weight of polystyrene blocks, 23,500; overall molecular weight 107,000.

Moisture and oxygen must be rigorously excluded from the system if the synthesis is to be successful. Inadvertant introduction of trace protic impurities of metal oxides during the synthesis leads to the formation of polydimethylsilioxane homopolymer. This material, present in even small amounts, is detrimental to the mechanical properties of the final block copolymer. This effect can be illustrated by deliberately including known amounts of polydimethylsiloxane in a block copolymer of demonstrated mechanical strength. These effects are summarized in Table I.

Structure–Property Relationships

The characteristics of the copolymers, which range from elastomers to low modulus thermoplastics of varying mechanical and rheological properties, depend on molecular parameters that can be predetermined in the synthesis by the relative amounts of monomers, initiator, and coupling reagent (R_2SiX_2) used. These parameters are overall molecular weight, the relative and absolute sizes of the component blocks, and the glass transition temperature of the "hard" blocks. The influence of each of these on properties is considered separately.

The effect of overall molecular weight on mechanical and rheological properties was determined by measuring properties on samples obtained by fractional precipitation of two different $(BAB)_x$ block copolymers. One copolymer containing 30 wt % polystyrene gave seven fractions where x varied from 1.6 to 60. The silicone content was nearly identical between fractions. A second copolymer containing 50 wt % polystyrene gave five fractions. Tensile properties were measured on solution cast films from the first series. The data showed that x must be greater than 2 for the films to show any significant mechanical strength. Tensile strength increases sharply with molecular weight until x reaches 8 to 10, after which little further increase is seen. The general shape of the stress–strain curve is essentially the same for each fraction. The envelope of the stress–strain curves and points of failure are shown in Figure 1.

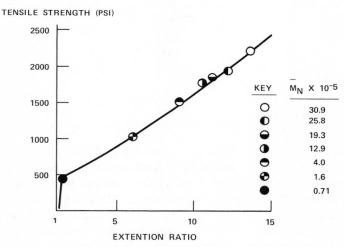

Figure 1. Effect of molecular weight on tensile properties of fractionaly precipitated samples of (BAB)_x which contain 30 wt % polystyrene and $\overline{M}_A = 13,500$

The effect of overall molecular weight or the number of blocks on rheological properties for the samples from the second fractionation can be illustrated as a plot of reduced viscosity *vs.* a function proportional to the principal molecular relaxation time (Figure 2). This function includes the variables of zero shear viscosity, shear rate, γ, and absolute temperature, T, in addition to molecular weight, and allows the data to be expressed as a single master curve (*10*). All but one of the fractions from the copolymer containing 50% polystyrene fall on this

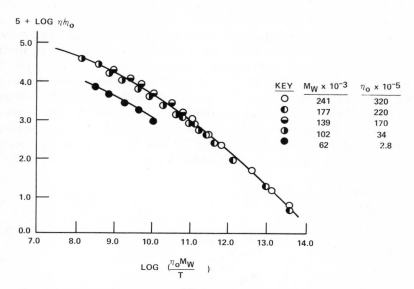

Figure 2.   Effect of molecular weight on reduced bulk viscosity expressed as a master curve for fractionally precipitated samples of $(BAB)_x$ which contain 50 wt % polystyrene and $\overline{M}_A = 13,500$. $T = 463°K$.

curve.   The exception is the fraction of lowest molecular weight which contains a high proportion of the species $(BAB)_x$ where $x = 1$.   The value of -0.99 for the limiting slope of the master curve in Figure 2 is unusual.   This is substantially higher than the value of -0.82 predicted from theoretical considerations and the values observed for polystyrene and polydimethylsiloxane (*11*).

The molecular weight of the polystyrene blocks is determined by the ratio of monomer to initiator used in the synthesis and is critical in determining mechanical and rheological properties.   The data in Table II indicate that a block size of about 8000 is required to obtain useful

Table II.   Effect of Polystyrene Block Size on Mechanical Properties of Compression Molded Polystyrene–Polydimethylsiloxane Block Copolymers Containing 30% Polystyrene

$\overline{M}_n$ Polystyrene Block	Degree of Condensation, x	Ultimate Stress, psi	Ultimate Strain, %
4,000	3.3	240	120
7,700	3.6	700	260
11,100	3.9	950	550
12,300	3.5	1,020	350
13,550	3.3	1,030	480

mechanical properties and that a molecular weight greater than 12,000 gives little further improvement.

Melt viscosity increases as the molecular weight of the polystyrene blocks is increased, but the effect tends to diminish as the rate of shear is increased. The influence of block size is expressed as a family of converging viscosity—shear rate curves for three copolymers of differing block size, (Figure 3). These curves also illustrate the non-Newtonian character of the polystyrene–polydimethylsiloxane block copolymers. The effect of changing block size cannot be expressed as a single master curve as in the case of overall molecular weight. Such master curves must be based on polymers of constant block size.

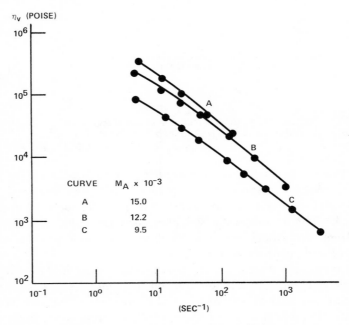

Figure 3. Effect of size of polystyrene blocks on true apparent viscosity at 190°C for a multiblock copolymer containing 40 wt % polystyrene; overall molecular weight = $130 \pm 10 \times 10^3$; $\overline{M}_A = 9.5 \times 10^3$.

The relative amounts of the two monomers used in the synthesis determine the relative size of the two blocks or the composition of $(BAB)_x$. This in turn determines whether mechanical behavior resembles that of thermoplastics or thermoplastic elastomers. The greater the polystyrene content, the greater the initial modulus and yield point of the block copolymer. Overall composition thus tends to dominate the parameters discussed above. At a given level of polystyrene the

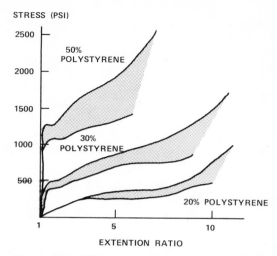

Figure 4. Effect of polystyrene content on tensile properties of polystyrene–polydimethylsiloxane multiblock copolymers

other parameters change ultimate mechanical properties but effect only minor alterations in the general shape of the stress–strain curve. Envelopes of stress–strain curves where overall molecular weights and the size of polystyrene blocks change broadly at three given contents of polystyrene are illustrated in Figure 4.

The amount of polystyrene also influences the permeability of the block copolymers to gases. Thus, a compression molded film of thermoplastic elastomer containing 20 wt % polystyrene shows a permeability towards oxygen typical of a silica-filled silicone elastomer, 49.2×10^{-9} cm^3-cm/cm^2-sec, cm Hg at 25°C. The permeability of films with increasing amounts of polystyrene rapidly decreases linearly to an inflection point at 50 wt % polystyrene where the permeability is 3.6×10^{-9} cm^3-cm/cm^2-sec, cm Hg. A similar trend is noted with nitrogen and CO_2. The inflection point at 50 wt % might be a consequence of the much less permeable polystyrene predominating in the continuous phase of the microdisperse two-phase system (3). The method of fabrication can also be an be an important factor in determining permeability.

Block copolymers constituted so that the hard A blocks show increased glass temperatures might be expected to show better ultimate tensile properties than comparable block copolymers of a lower T_g in the glassy phase (12). This is demonstrated in the present system by substituting the major portion of the polystyrene in the present system for poly(α-methylstyrene). A short length of polystyrene is included at each chain end of the hard block to facilitate the second step of the synthesis and to give a more stable polymer. The effect of replacing

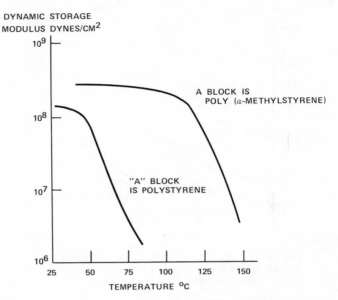

Figure 5. Effect of T_g in the glassy block on thermo-mechanical properties of $(BAB)_x$ containing 40 wt % "A"; curve A, "A" block is poly(α-methylstyrene); curve B, "A" block is polystyrene.

Table III.   Effect of T_g of the Hard "A" Block on Tensile Properties of $(BAB)_x$ Containing 40 wt % of A

Temperature, °C	A = Polystyrene		A = Poly(α-methylstyrene)	
	Tensile at Break, psi	Elongation at Break, %	Tensile at Break, psi	Elongation at Break, %
25	1550	800	2400	700
50	1080	1000	—	—
100	90	300	870	800
130	—	—	300	1160
150	—	—	90	1100

polystyrene with poly(α-methylstyrene) in the A blocks on thermo-mechanical properties is summarized in Figure 5 and Table III.   Figure 5 shows that the modulus of the material based on poly(α-methylstyrene) is retained to temperatures 70° greater than a corresponding material based on polystyrene.   Table III shows a significantly greater tensile strength at a given temperature for copolymers based on poly(α-methylstyrene).

Multiblock copolymers based on poly(α-methylstyrene) also show significantly better oxidative thermal stability than the block copolymers based on polystyrene. Thus, polystyrene–polydimethyldisiloxane multiblock copolymers lose half of their tensile strength after 80 hours with considerable yellowing at 150°C in air, but corresponding materials based on poly(α-methylstyrene) show no discoloration or loss in tensile properties under the same conditions.

Literature Cited

1. Merker, R. L., Scott, M. M., *J. Polymer Sci.* (1964) **2**, 15.
2. Vaughn, H. A., *J. Polymer Sci., Pt. B* (1969) **7**, 569.
3. Saam, J. C., Fearon, F. W. G., *Ind. Eng. Chem., Prod. Res. Develop.* (1971) **10**, 10; *Polymer Preprints* (1970) **11**, (2), 455.
4. Dean, J. W., *J. Polymer Sci., Pt. B* (1970) **8**, 677.
5. Noshay, A., Matzner, M., Merriam, C. N., *Polymer Preprints* (1971) **12** (1) 247.
6. Greber, G., Metzinger, L., *Makromol. Chem.* (1960) **39**, 167.
7. Morton, M., Rembaum, A. A., Bostick, E. E., *J. Appl. Polymer Sci.,* (1964) **8**, 2707.
8. Saam, J. C., Gordon, D. J., Lindsey, S. L., *Macromolecules* (1970) **3**, 1.
9. Gilman, H., Carteledge, F. K., *J. Organometal Chem.* (1964) **2**, 447.
10. Graessly, W. N., *J. Chem. Phys.* (1967) **47**, 1942.
11. Lee, C. L., Polmanteer, K. E., King, E. G., *J. Polymer Sci., Pt. A-2* (1970) **8**, 1909.
12. Fetters, T. J., Morton, M., *Macromolecules* (1969) **2**, (5), 453.

RECEIVED May 1, 1972.

16

Cationic Polymerization of Cyclic Ethers Initiated by Macromolecular Dioxolenium Salts

YUYA YAMASHITA

Nagoya University, Nagoya, Japan

Cationic polymerization of tetrahydrofuran (THF) with 2-methyl-1,3-dioxolenium perchlorate occurs without chain transfer and termination, and the initiation reaction proceeds by bonding. Polystyrene with terminal dioxolenium groups was synthesized from polystyrene anion using ethylene oxide, adipoyl chloride, ethylene bromohydrin, and silver perchlorate. Polymerization of THF by polystyrene dioxolenium salt in 1-nitropropane yielded a block copolymer of styrene and THF. In addition, poly-THF with terminal dioxolenium groups was synthesized from polytetramethylene glycol. Polymerization of 3,3-bischloromethyloxetane with poly-THF yielded a block copolymer; no chain transfer but some termination took place. Polymerization of 7-oxa[2.2.1]bicycloheptane, also with poly-THF, yielded homopolymer as well as a block copolymer; significant chain transfer was observed. Polymerization of dioxolane and tetraoxane by the same initiator yielded no block copolymer, showing the different mechanism of the initiation reaction with cyclic formals.

In recent years, several active catalysts have been reported for the cationic polymerization of tetrahydrofuran (THF); the reaction was shown to proceed with little termination or chain transfer (3). Several attempts have been made to prepare block copolymers by using the "living nature" of the cationic polymerization of THF (1, 2, 4, 5). Polymerization of THF and other cyclic ethers by macromolecular initiators can provide an interesting procedure for synthesizing block copolymers consisting of well-characterized sequences. Although polymers having terminal groups such as acyl halide or epoxide can initiate THF polymerization in the presence of a Lewis acid, some transfer or termination reactions occur leading to insufficient block copolymers (7). The 2-substituted dioxolenium group with perchlorate anion has been shown to have excellent initiating efficiency, and practically no transfer or termination reaction was observed in the polymerization of THF (7). This paper describes the preparation of polystyrene and poly-THF with dioxolenium end groups. The polymerization of THF and other cyclic ethers was studied in a high-vacuum system.

Experimental

All reagents and solvents were dried and distilled or recrystallized in a vacuum system.

Preparation of Macromolecular Dioxolenium Salts. "Living" polystyrene prepared by the polymerization of styrene in THF with α-methylstyrene tetramer dianion reacted with a 2.1-molar amount of ethylene oxide for three hours at room temperature; a 6.6-molar amount of adipoyl chloride was added, and the mixture was stirred for 20 hours; a 20-molar amount of ethylenebromohydrin was added. This mixture was stirred for 44 hours. The bromoethylated polystyrene was precipitated in excess methanol and freeze-dried from benzene in a vacuum system. A 1-nitropropane solution of polystyrene dioxolenium salt was prepared by reaction of bromoethylated polystyrene with silver perchlorate in 1-nitropropane. Silver bromide was removed from the reaction mixture by filtration. Molecular weight of the product was measured by a vapor-pressure osmometer; it was 1910 for "living" polystyrene and 5190 for the bromoethylated polystyrene. Bromine analysis of the bromoethylated polystyrene showed 67.9% of the calculated value.

Polytetramethylene glycol of molecular weight 2000 was freeze-dried from benzene in a vacuum. The bromoethylated prepolymer was obtained by reaction of the glycol, first with adipoyl chloride, then with ethylene bromohydrin (Table II) and precipitation in water and n-hexane. Nitromethane solution of the poly-THF dioxolenium salt was prepared after filtration of silver bromide from the reaction product of bromoethylated poly-THF with silver perchlorate in nitromethane.

Polymerization Procedure and Characterization. Cyclic ethers and formals were polymerized by adding a measured amount of monomer into the initiator solution at 0°C. The polymer was precipitated with methanol or ethyl ether and freeze-dried from benzene or fractionated by chloroform. The block copolymer of styrene and tetrahydrofuran was dissolved in 1-butanol and refluxed for 12 hours with sodium metal. The solution was washed with water, and the 1-butanol was distilled off. The residual polymer was freeze-dried from benzene, and poly-THF was extracted with 2-propanol in a Soxhlet apparatus.

Results and Discussion

Polymerization of THF by Polystyrene Dioxolenium Salt. Polystyrene with terminal dioxolenium salt groups was prepared by the scheme shown here, where PSt means polystyrene segment containing two end groups:

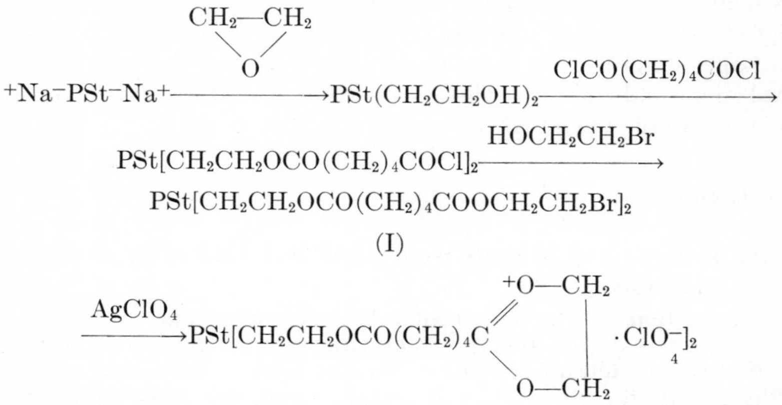

$$\begin{array}{c} CH_2{-}CH_2 \\ \diagdown \diagup \\ O \end{array}$$

$$^+Na^-PSt^-Na^+ \longrightarrow PSt(CH_2CH_2OH)_2 \xrightarrow{\;ClCO(CH_2)_4COCl\;}$$

$$PSt[CH_2CH_2OCO(CH_2)_4COCl]_2 \xrightarrow{\;HOCH_2CH_2Br\;}$$

$$PSt[CH_2CH_2OCO(CH_2)_4COOCH_2CH_2Br]_2$$

$$(I)$$

$$\xrightarrow{\;AgClO_4\;} PSt\Big[CH_2CH_2OCO(CH_2)_4C \begin{array}{c} {}^+O{-}CH_2 \\ \diagup\!\!\diagup \\ \\ \diagdown \\ O{-}CH_2 \end{array} \cdot ClO^-\Big]_{2}$$

The polymerization of THF with polystyrene dioxolenium salt was carried out in 1-nitropropane at 0°C. Table I shows the increase in molecular weight with conversion. The resulting copolymer was a

Table I. Polymerization of THF by PSt–dioxolenium Salt

Number	*1*	*2*	*3*	*4*	*5*
Initiator [a] (grams)	1.663	1.298	0.995	0.558	0.291
THF (ml)	66.3	51.7	39.7	21.4	11.6
1-Nitropropane (ml)	14.6	11.4	8.7	4.7	2.6
Polymerization time [b] (hr)	22	38	57	77	162
Polymer (grams)	6.90	9.31	8.81	6.32	5.50
Conversion of THF (%)	8.9	17.5	22.2	30.5	50.7
$\overline{M}_n \times 10^{-3}$ observed	63.8	109	161	191	437
$\overline{M}_n \times 10^{-3}$ of poly-THF after alcoholysis					
Observed	54	68	103	125	219
Calculated	12.1	23.7	30.1	41.2	68.4

[a] Initiator concentration = 2.65×10^{-3} moles/liter
[b] At 0°C, [THF] = 10.1 moles/liter

block copolymer of the THF-St-THF type, and free from homopolymers. The poly-THF segment of the block copolymer was isolated by alcoholysis. Decrease of the molecular weight by chain cleavage proved formation of the block copolymer. However, the molecular weight of the poly-THF segment was higher than the calculated value based on the assumption of the formation of one polymer chain from every dioxo-

lenium group, thus showing the slow initiation from the stable dioxolenium group.

Polymerization of THF by Poly-THF Dioxolenium Salt. Poly-THF
with terminal dioxolenium salt groups was prepared by this scheme,
where PTHF means polytetrahydrofuran segment containing two end
groups:

$$HO[(CH_2)_4O]_nH \xrightarrow{ClCO(CH_2)_4COCl}$$

$$PTHF[OCO(CH_2)_4COCl]_2 \xrightarrow{HOCH_2CH_2Br}$$

$$PTHF[OCO(CH_2)_4COOCH_2CH_2Br]_2 \xrightarrow{AgClO_4}$$

$$PTHF[OCO(CH_2)_4C \overset{+O—CH_2}{\underset{O—CH_2}{\diagup\diagdown}} \cdot ClO_4^-]_2$$

(II)

The bromoethylated prepolymer (II) was prepared and analyzed
(Table II). The prepolymer reacted with silver perchlorate in nitromethane. The precipitated silver bromide was weighed to measure
completion of the reaction.

Table II.   Preparation of Prepolymer [II] from PTG
$$BrCH_2CH_2OCO(CH_2)_4CO[O(CH_2)_4]_{27.6}OCO(CH_2)_4COOCH_2CH_2Br$$

Number	*2*	*3*	*4*
Reagents [a]			
PTG 2000 (10^{-3} mole)	2.94	2.17	5.10
Adipoyl chloride (10^{-3} mole)	9.02	5.96	12.5
Reagents [a]			
Ethylene bromohydrin (10^{-3} mole)	10.08	7.04	13.1
Solvent (ml)	THF 20	EDC 20	EDC 50
Pyridine (ml)	1.5	1.0	—
Prepolymer			
$\overline{M}_n$			
Observed	2510	—	1900
Calculated	2470	—	2470
Number of ester groups			
per polymer chain [b]	4.19	3.93	3.56

[a] Reaction of PTG and adipoyl chloride was carried out for 12 hours at room temperature; further reaction with ethylene bromohydrin was for 12 hours.
[b] The calculated number of ester groups per polymer chain was four; the observed
value was measured from IR spectra (7).

The polymerization of THF with poly-THF dioxolenium salt was carried out in nitromethane at 0°C. The results are shown in Table III. The agreement of the observed $\overline{M}_n$ with the calculated value from conversion, assuming the formation of "living" polymer was fairly good. The number of ester groups per polymer chain analyzed from IR spectra (7) remains four, as shown in Table III; the values are similar to those in Table II. These results substantiate the formation of macromolecular dioxolenium salt at both ends of PTHF.

Table III.   Polymerization of THF by PTHF-Dioxolenium Salt

Number	4-1	2-1
Prepolymer [II] (10^{-4} mole)	9.63	11.63
CH_3NO_2 [a] (ml)	4.57	9.05
THF (10^{-2} mole)	6.10	12.08
Polymerization time [b] (hr)	12	15
Polymer (grams)	2.99	5.82
Conversion of THF (%)	26.4	21.8
$\overline{M}_n$		
Observed	3400	3500
Calculated [c]	3100	3700
Number of ester groups per polymer chain [d]	3.86	3.70

[a] The initiator was prepared by reaction of the prepolymer (Table I) with silver perchlorate for three hours at 0°C in CH_3NO_2.
[b] At 0°C
[c] From conversion
[d] From IR spectra (7)

Polymerization of Cyclic Ethers and Formals by Poly-THF Dioxolenium Salt. The polymerization of cyclic ethers and formals by PTHF-dioxolenium salt was carried out to clarify the presence of termination or transfer reactions. The results are shown in Table IV. In the polymerization of 3,3-bischloromethyloxetane (BCMO), block copolymer soluble in chloroform and having the expected molecular weight was formed; the homopolymer of BCMO insoluble in chloroform was not observed. The block copolymer showed crystalline bands of BCMO at 700, 860, and 890 cm⁻¹, suggesting the formation of ABA block.

The polymerization of BCMO by dioxolenium salt seems to involve the initiation by bonding and the "living" nature of the propagation without chain transfer. However, attempted polymerization of BCMO by low concentrations of initiator ($10^{-5}M$) stopped at low conversion of BCMO after several days, showing the importance of the termination reaction (6).

In the polymerization of 7-oxa[2.2.1]bicycloheptane (OBH), the formation of block copolymer soluble in chloroform and homopolymer

Table IV.   Polymerization of Cyclic Ethers and Formals by PTHF-dioxolenium Salt

Number	4–2	4–3	2–2	4–4	3–1
Comonomer	BCMO	OBH	OBH	TEX	DOL
PTG initiator (10^{-4} mole)	9.95	9.37	10.40	10.05	5.04
Comonomer (10^{-2} mole)	2.99	4.45	6.67	3.17	10.9
CH_3NO_2 (ml)	8.10	4.36	6.98	21.9	10.0
Polymerization time [a] (hr)	14.5	0.67	24	0.17	1
Block Copolymer [b] (grams)	2.99	1.06	5.61	—	1.65 [c]
MeOH-Soluble Polymer (grams)	0.185	0.143	0.05	1.80	
$CHCl_3$-Insoluble Polymer (grams)	0	1.52	2.21	1.03	2.09 [d]
Conversion of Comonomer (%)	25.6	21.6	46.0		38.2
Block Copolymer $\overline{M}_n$					
Observed	3500	1500	2650	—	2600
Calculated	3100	—	—	—	—
Comonomer content (mole %)	42.0	—	21.6	—	29.7

[a] At 0°C
[b] MeOH-insoluble, $CHCl_3$-soluble part
[c] Soluble in hot petroleum ether
[d] Insoluble in hot petroleum ether, soluble in $CHCl_3$ and benzene, DOL homopolymer

of OBH insoluble in chloroform was observed.   Thus, although dioxolenium salt can initiate OBH polymerization by bonding, considerable chain transfer to monomer seems unavoidable.

In the polymerization of tetraoxane (TEX), we did not observe the formation of block copolymer, thus showing a different mechanism in the intiation reaction.   This is also true for the polymerization of 1,3-dioxolane (DOL).   The formation of DOL homopolymer suggests the presence of a chain-transfer reaction.   Also, the presence of homopolymer mixtures suggests that initiation by bonding was not complete.

Conclusions

Polymerization of cyclic ethers by macromolecular dioxolenium salt yielded block copolymers.   No termination or chain transfer was observed with THF, but some termination with BCMO and considerable chain transfer with OBH occurred.   Cyclic formals such as TEX or DOL do not form block copolymers, showing that the initiation mechanism is not *via* bonding.

Literature Cited

1. Aoki, S., Otsu, T., Imoto, M., *Kogyo Kagaku Zasshi* (1964) **67**, 971, 1953, 1955; *J. Polym. Sci.* (1964) **B2**, 223.
2. Berger, G., Levy, M., Vofsi, D., *J. Polym. Sci.* (1966) **B4**, 183.
3. Dreyfuss, P., Dreyfuss, M. P., *Advan. Polym. Sci.* (1967) **4**, 528.
4. Lambert, J. L., Goethals, E. J., *Makromol. Chem.* (1970) **133**, 289.
5. Saegusa, T., Matsumoto, S., Hashimoto, Y., *Macromolecules* (1970) **3**, 377.
6. Saegusa, T., Matsumoto, S., Hashimoto, Y., *Macromolecules* (1971) **4**, 1.
7. Yamashita, Y., Kozawa, S., Hirota, M., Chiba, K., Matsui, H., Hirao, A., Kodama, M., Ito, K. *Makromol. Chem.* (1971) **142**, 171.

RECEIVED April 1, 1972.

Photopolymerization in the Organic Crystalline State

G. WEGNER

Institut für Physikalische Chemie der Universität Mainz, II Ordinariat, Mainz, Germany

Organic solid state photopolymerizations have become a powerful method of polymer synthesis and are interesting for substantiating optical data. Recent advances are described under several aspects. One is the increasing understanding of the relationship between molecular structure, monomer packing properties, and lattice-controlled photoreactivity (topochemistry). Examples are taken from Hasegawa's four-center type photopolymerization and from the topochemical polymerizations of monomers with conjugated triple bonds. Solid state reactions are considered to be influenced by the phase properties of monomer–polymer systems and by lattice defects. These reactions can be treated kinetically as special solid–solid transformations. The four-center type photopolymerization seems to be a two-phase reaction; diacetylene polymerization is one-phase. Both types proceed via addition of photoexited monomer molecules to chain ends.

The term "molecular engineering" has been used to characterize the recent advances in solid state photochemistry (*1*). This rapidly growing area is interesting because of the inherent possibilities for using organic solid state reactions as a means of reversible or non-reversible data fixation. This is obvious from the fact that we know of numerous examples of organic molecular crystals which undergo lattice-controlled photoreactions or photopolymerizations. Although the search for new systems for substantiating optical information may be a major driving force in this area, one may not forget that lattice-controlled photoreactions were developed from a mere scientific curiosity to a promising tool of preparative polymer chemistry.

General Scope of Solid State Photoreactions

The term lattice control describes the fact that the monomer molecules are already arranged within the lattice of the organic photoreactive crystal in a geometry which favors a special addition or polymerization. In ideal cases, adjacent molecules are already arranged with respect to each other so as to preform the geometry of the activated complex of a bimolecular photoreaction. On the other hand, a wealth of data on the relationship between molecular structure and packing properties of organic molecules is at our disposal mainly through the work of Kitaigorodskii (2) and G. M. J. Schmidt (3). This enables us—at least in principle—to "design" lattices of photoreactive molecules, thus providing an extremely selective method of photoreaction by lattice control. This is the basic idea behind organic topochemistry.

Although this concept looks fairly straightforward and has been successfully applied (1, 4), the kinetics of these solid state reactions impose severe restrictions when generally applied. First, lattice defects and crystal texture obviously play a very important role either as locations where the reactions start or finish or as centers where chain transfer or irregular reactions may occur. (5). Secondly, during the reaction a second type of molecule is formed—namely the photopolymer which has a different molecular structure and therefore, in general, has different packing properties. As a result, phase separation may take place according to the reaction temperature to an extent which is given by the phase diagram of the binary system monomer–polymer (6). Depending on the type of phase diagram, a liquid monomer–polymer mixture will be formed, thus destroying the original lattice. Even if no phase separation takes place, relaxation processes of the newly formed macromolecules may take place because of a misfit between the molecular geometry of the photopolymer and the monomer lattice; this increases the number of defects in the monomer lattice and the probability for unwanted side reactions such as crosslinking (7). Sometimes these difficulties can be circumvented by lowering the reaction temperature so that the reaction product is frozen in and no phase separation or relaxation can take place because of the lack of molecular mobility. One other method is to attach side groups to the photoreactive system which do not change the basic geometry of the lattice but only serve to increase the stability of the lattice—that is to increase the melting point of the system.

Molecular Structure and Solid State Reactivity

To show solid-state reactivity the molecular structure must be such as to allow packing of the molecules with close contact between the photoreactive groups. According to G. M. J. Schmidt (*8*) a minimum distance of 4 A is required to achieve a four-center type photoaddition of double bonds in the solid state, and it seems that this is a general rule for all kinds of solid state photoreactions.

Among the numerous examples of organic solid state photopolymerizations (*e.g.*, *9-12*) only two systems have been thoroughly studied with regard to the relationship between crystal and molecular structure. The first is the so called four-center type photopolymerization (Reaction 1) which was investigated by Hasegawa and co-workers (*13*); this is a straightforward translation of the work of G. M. J. Schmidt (*3*) and his students on the photodimerization of cinnamic acids in the crystalline state into polymer chemistry.

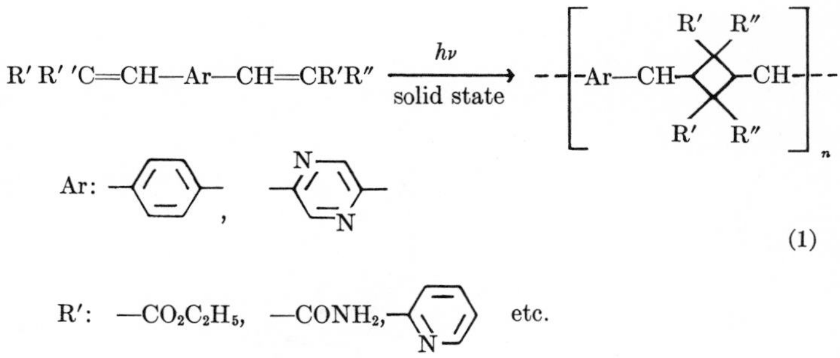
R′R′′C=CH—Ar—CH=CR′R″
hν
solid state
Ar—CH
CH
R′ R″
R′ R″
n
Ar:
N
N
R′: —CO₂C₂H₅, —CONH₂,
N
etc.

(1)

R″: H, CN

The second system is the topochemical polymerization of monomers with conjugated triple bonds according to Reaction 2 (*4, 14*).

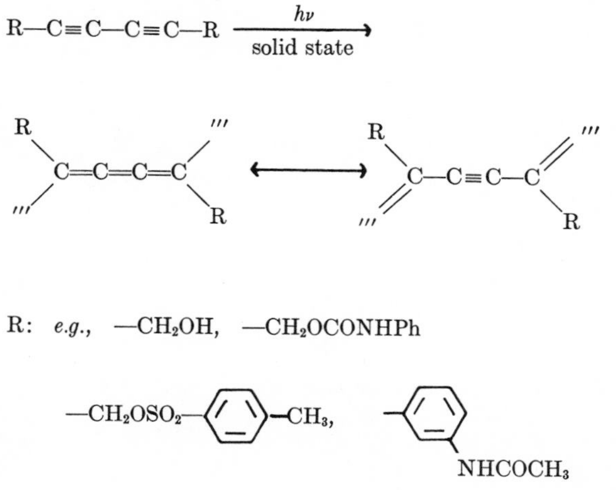
R—C≡C—C≡C—R
hν
solid state
R
C=C=C=C
R
R
C—C≡C—C
R
R: e.g., —CH₂OH, —CH₂OCONHPh
—CH₂OSO₂—
—CH₃,
NHCOCH₃

(2)

According to the rules developed by Schmidt two types of photo-reactive arrangements of trans-substituted olefins are possible in the solid state. A head-to-tail arangement of the monomers (in crystals with a center of symmetry) leads always to dimers with a stereochemical arrangement at the cyclobutane ring corresponding to α-truxillic acid in the case of dimerization of cinnamic acid. This type of polymerization occurs with crystals of 2,5-distyrylpyrazine (I) (*15*), 1,4-bis[β-pyridyl-(2)-vinyl]benzene (III) (*16*) and *p*-phenylene diacrylic acid dimethyl ester (IV) (*17*) as demonstrated by x-ray analysis of the corresponding monomer and polymer lattices. Polymers of very high molecular weight are obtained according to Reaction 3. Type II polymers exhibit one cyclobutane ring per base unit with 1,3-trans configuration of the substituents. Such polymers cannot be obtained by the usual synthetic methods of polymer chemistry in solution.

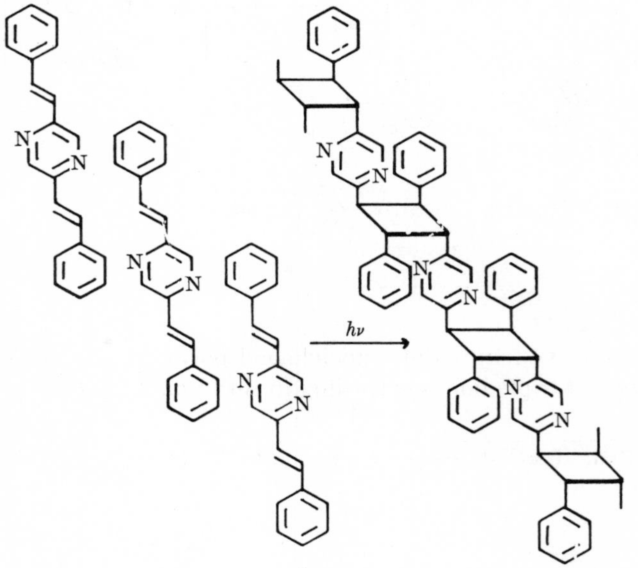

(3)

I   II

III

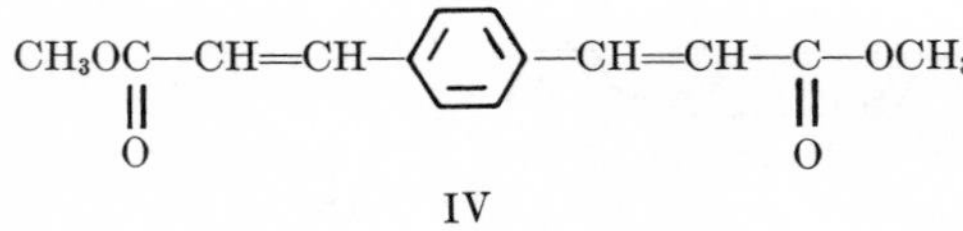

IV

The second type of photoreactive arrangement of monomer molecules in crystals with mirror symmetry leads to polymers with a stereochemistry at the cyclobutane ring corresponding to β-truxinic acid as shown in Reaction 4. So far, crystallographic evidence for this type of polymerization is lacking but should be coming in near future.

Recent work of Green and Schmidt (1) shows how the packing can be influenced to obtain either of the two reaction products. As they have reported, derivatives of 2,4-dichlorocinnamic acid always pack in a head-to-head arrangement, giving rise to derivatives of β-truxinic acid. Thus, polymers with the same stereochemistry as V (Reaction 4) should be produced by using the 2,4-dichlorophenyl residue as the guiding group.

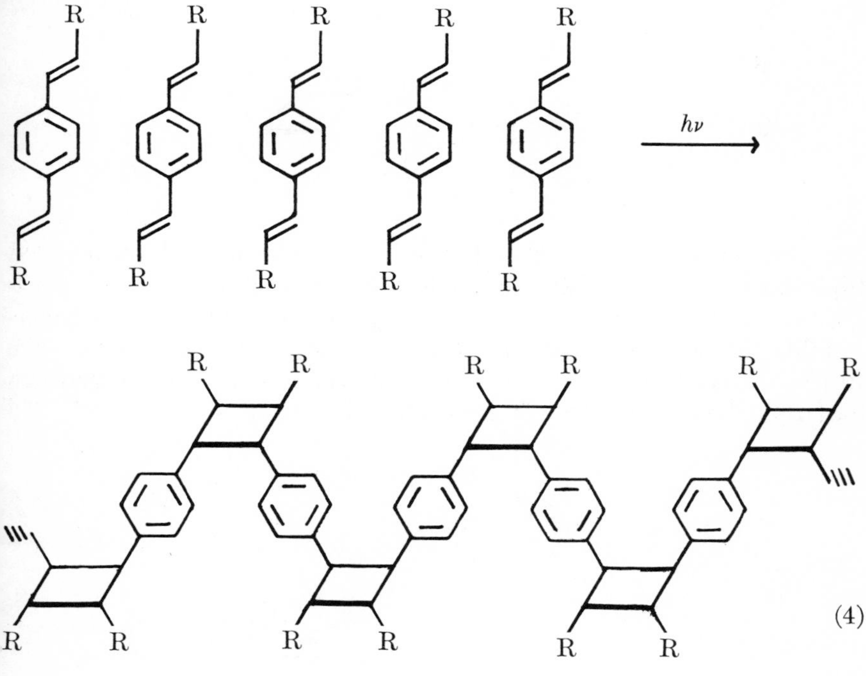

V

Further speculation along these lines led to the prediction that it should be possible to produce optically active polymers by the four-center type reaction if one uses a type VI monomer where $R \neq R'$; if one succeeds, this monomer will crystallize in a space group with polar axis (*e.g.*, P1) to give a polymer of type VII (Reaction 5) from enantiomorphous crystals. In fact, the production of optically active reaction products by solid state method has already been predicted in more general terms by H. Morawetz in 1966 (*32*).

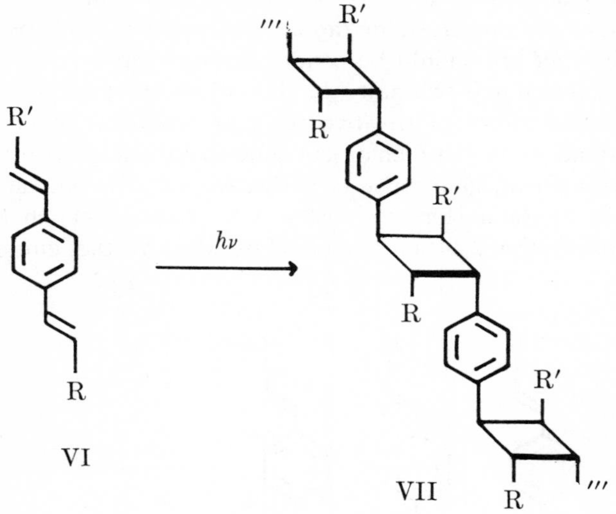

$$(5)$$

The topochemical polymerization of monomers with conjugated triple bonds is an interesting example of the relationship between packing properties and reactivity in the solid state.   For solid state polymerizability the monomer molecules must be arranged in a herringbone-like structure according to Figure 1 so that only one nearest distance between the triple bonds of adjacent molecules of 3 to 4 A is affected.   To

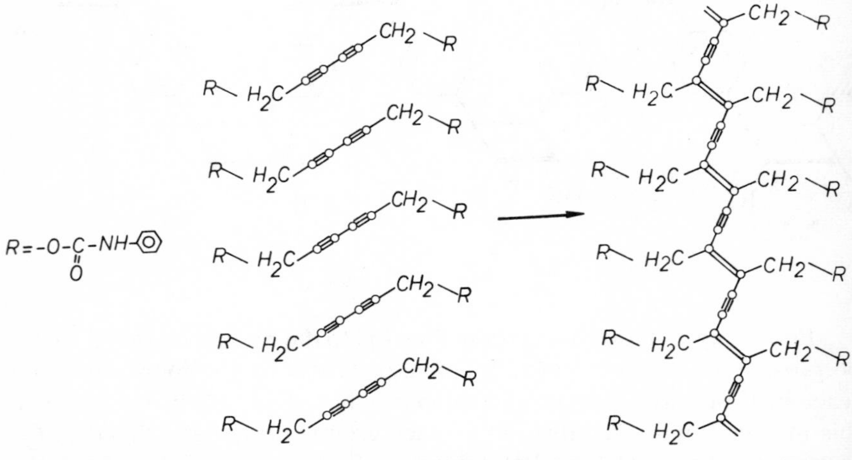

Figure 1.   Scheme of the topochemical polymerization of monomers with conjugated triple bonds

achieve that, adjacent molecules must be related by a monclinic displacement (2). The amount of monoclinic displacement is governed by the substituents at the conjugated triple bonds which also stabilize the lattice and prevent phase separation during polymerization. The best results are obtained with substituents ready to form hydrogen bonds. Hydrogen bonding leads to directional bonding with monoclinic displacement and also provides for very rigid lattices. The reaction is called a shearing reaction because the type of molecular mobility which is needed to bring about polymerization is similar to a shearing action (14, 22, 25).

Crystallographic analysis confirmed (18, 19, 20, 25) that in the case of the topochemical polymerizations according to Figure 1 monomer and polymer lattice are essentially isomorphous, and polymers are formed with polyconjugated backbone and substituents in the trans position with respect to the double bond.

Recent work in our laboratory showed that the reaction can be extended to higher conjugated acetylenes. There is substantial evidence to assume that the solid state photopolymerization of octatriin-1,6-diol (VIII) leads not to polymer IX with a polyconjugated backbone and two conjugated triple bonds per base unit but to the "normal" polymerization product X (22).

$$HO—CH_2—(C≡C)_3—CH_2—OH$$

VIII

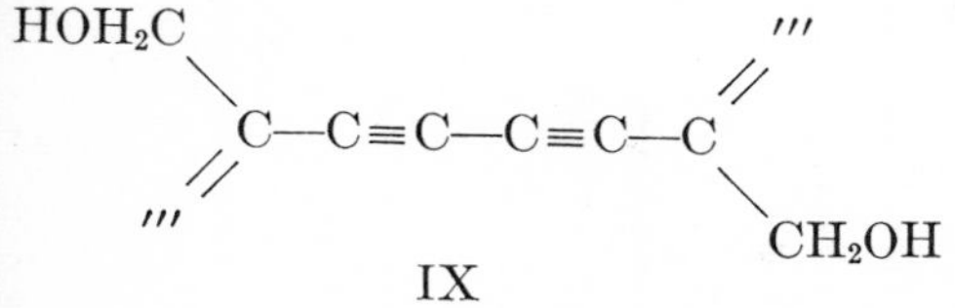

$$HOH_2C \diagdown \overset{///}{\underset{///}{C}} —C≡C—C≡C—\overset{///}{\underset{\diagdown CH_2OH}{C}}$$

IX

$$HOH_2C \diagdown \overset{///}{\underset{///}{C}} —C≡C—\overset{///}{\underset{\diagdown C≡C—CH_2OH}{C}}$$

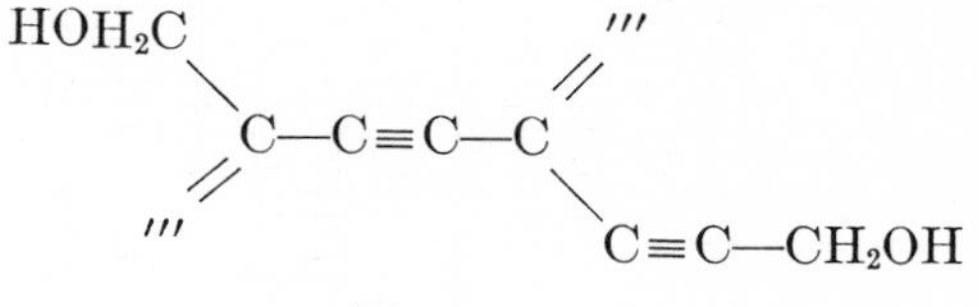

X

Kinetics of Solid State Photopolymerizations

Kinetics of Phase Change. Two principally different mechanisms of solid state reactions may be visualized if one treats these reactions as special solid–solid transitions.   The first mechanism often encountered in high energy radiation-induced as well as in light-induced polymerizations is a two-phase mechanism.   The first step of the reaction is the formation of nuclei of the polymer phase, and polymer chain growth is intimately connected with the growth of the second phase as a morphological and crystallographic entity.   One obtains domains of polymer or oligomer molecules at first and small polymer crystallites at higher conversion which are oriented within the monomer crystal matrix.   This situation is illustrated in Figure 2a.   The general shape of the monomer crystal may be retained at higher conversion, but fibrillization takes place if there are considerable differences in lattice parameters between monomer and polymer phase.   This is true in many cases for Hasegawa's four-center type photopolymerization (15, 16, 23, 24, 25).   The formation of the second phase can be conveniently followed by x-ray methods from the

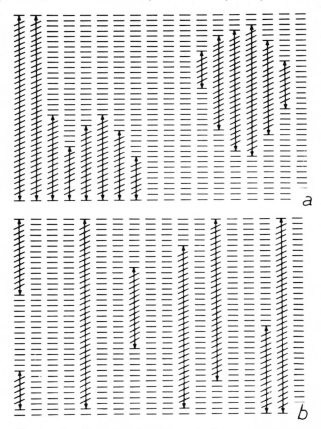

Figure 2.   Two mechanisms of the transition from monomer to polymer phase; (a) heterogeneous, (b) homogeneous growth of polymer chains.

appearance of new reflections from the beginning (20, 26) or by polarizing microscopy because monomer and polymer phase show different optical properties.   Often, the whole monomer crystal is broken down into numerous small crystallites which seem to be misoriented with respect to each other and the former lattice.   This generally occurs in the solid state dimerization of cinnamic acid derivatives.

It is important to recognize that in such cases the rate law of polymerization as obtained by determination of the polymer yield reflects the growth of morphological entities rather then the formation of molecular intermediates and may be determined by the rate of nucleation and by the rate of diffusion of the monomer molecules across the interphase between monomer and polymer lattice.   This is well established in the somewhat related case of radiation-induced polymerization of crystalline trioxane (27, 28).

The second mechanism of phase change is that of a one-phase reaction.   The polymer chains are formed at random within the monomer lattice—that is, arrays of monomer molecules are transformed into single polymer chains as shown in Figure 2b.   Polymerization is a kind of isomorphous replacement of monomer units by polymer chains.   Therefore, at any degree of conversion the polymerizing crystal is a type of solid solution of polymer chains extended along definite crystallographic axis and starting from points distributed at random throughout the monomer lattice.   It is a kinetically forced isomorphism of monomer and polymer in the same lattice.

The latter mechanism is observed in the polymerization of diacetylenes, and it is the true reason why large single crystals with extended chain morphology can be obtained from these monomers.   Experimentally, evidence for the one-phase mechanism is shown by x-ray scattering from the fact that new layer lines or new reflections do not appear during polymerization of a single crystal and that the lattice parameters of monomer and polymer are essentially the same (20, 25, 29).   In addition, a polymerizing single crystal always behaves as a single crystal over the whole range of conversion with respect to polarized light.   In some cases relaxation of the newly formed macromolecules were observed.   During these relaxations the monomer–polymer lattice was destroyed, and polymerization stopped at about 20-40% conversion. Recently, R. H. Baughman (25) has studied these relaxation processes occuring during polymerization by detailed x-ray measurements on 2,4-hexadiindiol.   Similar work was done in our laboratory using modification II of the bis(phenylurethane) of 2,4-hexadiindiol (7, 20, 29). Most of the diacetylenes occur in different crystalline modifications exhibiting different solid state reactivity.   This was investigated in more detail for three different modifications of the bis(phenylurethane) of

2,4-hexadiindiol. An interesting type of molecular "engineering" is possible with these modifications. X-ray work showed that the polymers formed from the three different modifications have the same identity period in one chain direction (c axis). Thus, the same chemical structure of the main chain is formed, and the polymers differ only in their packing of side chains as is well known for other polymers showing polymorphism.

The different modifications are interconvertible by thermal treatment. For example, modification II (as obtained by crystallization from anisole) is converted to modification III by annealing at 130°C. This is a normal change in modification, and polycrystalline material of modification III (mp. 172°C) is obtained from a single crystal of modification II. However, if a single crystal of modification II is photopolymerized to less than 1% conversion before annealing, modification III is obtained after annealing, highly oriented with respect to the photopolymer and the lattice of modification II. The point is that the photopolymer of modification II serves as the site of nucleation for the formation of the other modification.

Mechanism of Chain Growth. During solid state photopolymerization not much is known about the mechanism of excitation, the fate of the photoexited molecule, and the chemical transformations following the photochemical event. G. M. J. Schmidt established that the lattice geometry and the packing properties of the molecules play an important role, but only recently have data become available which allow a better description of chain growth in the crystalline lattice. From the work of Hasegawa and co-workers it became clear that the four-center type photopolymerization is a step reaction. Depending somewhat on wavelength of irradiation, dimers are formed first which then become oligomers and so on (26, 30) according to Reaction 6. Each step is an elementary photochemical reaction so that the quantum yield of the overall polymerization is rather low—ca. 0.04–1.20, depending on monomer structure (16).

$$\text{DSP (I)} \xrightarrow{\lambda > 400 \text{ nm}} \text{DSP-Oligomers (XI)} \xrightarrow{340 \text{ nm}} \text{DSP-Polymer (II)}$$

Monomer

(6)

XI

In contrast, photopolymerization of diacetylenes seems to be a continuous growth of polymer chains by addition of photoexited monomer molecules to active chain ends. The nature of the growing chain ends is not yet known, but interesting information can probably be drawn from spectral data on polymerizing single crystals of the three different modifications of 2,4-hexadiindiolbis(phenylurethane) mentioned above. Typical data are given in Figure 3, showing the variation of the crystal spectra with time on irradiation at 330 nm (monochromatic light).

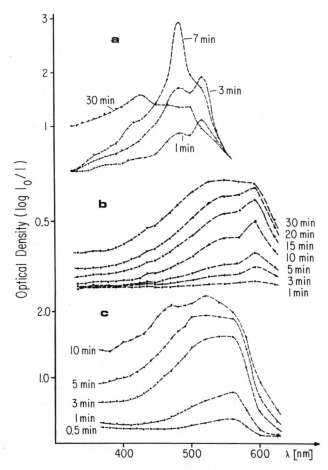

Figure 3. Spectra of single crystals of 2,4-hexadiindol-bis(phenylurethane) depending on irradiation time at 25°C. (a) Modification I (crystallized from dioxane water) irradiated at 330 nm with monochromatic light. (b) Modification II (crystallized from anisol) irradiated at 310 nm. (c) Modification III (from Modification II at 130°C) irradiated at 330 nm.

Figure 3a shows the spectra of a polymerizing crystal of modification I. At short irradiation times a strong absorption band with maximum extinction at 600 nm is formed. Soon after an absorption with very high extinction coefficient at 560 nm is formed with a simultaneous decrease of the 600-nm peak. Finally, the absorption peak of the polymer with a maximum at 490 nm is formed with a simultaneous decrease of the 560-nm peak.

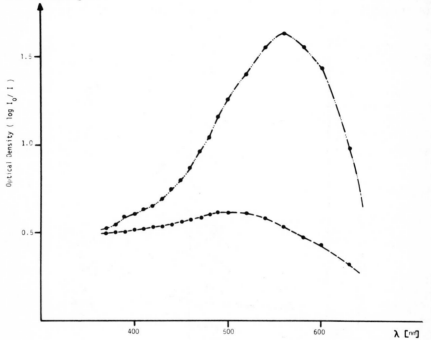

Figure 4. Spectra of a single crystal of Modification I of 2,4-hexadiindiolbis (phenylurethane) . . •. . *before extraction* —•— *after extraction Conversion ca. 2%.*

Similar behavior but with different rates of formation and interconversion of the various absorptions are found for the other two modifications of the same monomer (*cf.* Figure 3 b, c) (*31*). Extraction of partially polymerized crystals with a maximum absorption at either 600 or 560 nm leads to the immediate disappearance of these absorptions with only a small polymer absorption remaining (λ_{max} 490 nm) which corresponds to less than 2% conversion (*cf.* Figure 4). It must be assumed that the aforementioned absorptions result from the growing chain ends since the only products obtained after photopolymerization are a polymer of high viscosity and unchanged monomer. No indication was found for the presence of low molecular weight intermediates as checked by thin-layer chromatography.   The polymer chain ends may exist as either "dead" ends (λ_{max} 490 nm) or as "living" chain ends with a radical or carbene-like structure giving rise to the strong extinctions at 600 and 560 nm.   It may be further assumed that the species absorbing at the three different wavelengths are related by the following kinetic scheme.

$$\text{Monomer} \xrightarrow[\substack{330 \text{ nm}}]{h\nu} \text{Intermediate A} \rightarrow \text{Intermediate B} \rightarrow \text{"dead" polymer}$$

$$\qquad\qquad 330 \text{ nm} \quad (\lambda_{max}600 \text{ nm}) \qquad (\lambda_{max}560 \text{ nm}) \qquad (\lambda_{max}490 \text{ nm})$$

Further work to establish quantitatively the chemical nature of the intermediates and the quantum yield of the reaction is in progress.

Acknowledgment

The author thanks G. Nagy, Budapest, and B. Keggenhoff for helpful discussions.

Literature Cited

1. Green, B. S., Schmidt, G. M. J., IUPAC Congress, Boston 1971, OC 225.
2. Kitaigorodoskii, A. I., "Organic Chemical Crystallography," Consultants Bureau, New York, 1961.
3. Schmidt, G. M. J., in "Reactivity of the Photoexited Organic Molecule" pp. 227 ff, Wiley, New York, 1967.
4. Wegner, G., *J. Polymer Sci.* (1971) **B9** 133.
5. Thomas, J. M., *Advan. Catalysis* (1969) **19**, 293.
6. Hardy, Gy., *et al.* IIIrd., Tihany Symposium, 1971, B 15.
7. Wegner, G., Munoz-Escalona, A., Fischer, E. W., *Ber. Bunsenges. phys. Chem.* (1970) **74**, 909.
8. Cohen, M. D., Schmidt, G. M. J., Sonntag, F. I., *J. Chem. Soc.* (1964) 2000.
9. Yamada, M., Takase, J., Koutou, N., *J. Polymer Sci.* (1968) **B6** 883.
10. Bamford, C. H., Eastmond, G. C., Ward, J. C., *Proc. Roy. Soc. Ser. A* (1963) **271**, 357.
11. Matsuda, T., Higashimura, T., Okamura, S., *J. Macromol. Sci.* (1970) **A4** 1.
12. Ried, W., Wesselborg, K. H., *J. Prakt. Chem.* (1961) **12**, 306.
13. Suzuki, F., Suzuki, Y., Nakanishi, H., Hasewaga, M., *J. Polymer Sci., Al* (1969) **7**, 743, 2319.
14. Wegner, G., *Z. Naturforsch* (1969) **24b**, 824.
15. Sasada, Y., Shimanouchi, H., Nakanishi, H., Hasegawa, M., *Bull. Chem. Soc. (Japan)* (1971) **44**, 1262.
16. Nakanishi, H., Ueno, K., Hasegawa, M., *Chem. Letters,* in press.
17. Hasegawa M., private communication.
18. Hädicke, E., Mez, E. C., Krauch, C. H., Wegner, G., Kaiser, J., *Angew. Chem.* (1971) **83**, 253.
19. Hädicke, E., Penzien, K., Schnell, H. W., *Angew. Chem.* (1971) **83**, 1024.
20. Kaiser, J., Wegner, G., Fischer, E. W., *Israel J. Chem.* (1972) **10**, 157.

21. Wegner, G., *Makromol. Chem.* (1972) **154**, 35.
    Wegner, G., 23rd IUPAC Congress, Boston 1971, Macromolecular Preprints Vol. II, p. 818.
22. Kiji, H., Schulz, R. C., Kaiser, J., Wegner, G., *Polymer, in press.*
23. Nakanishi, H., Nakano, N., Hasegawa, M., *J. Polymer Sci.* (1970) **B8**, 755.
24. Hasegawa, M., *Chem. High Polymers (Japan)* (1970) **27**, 337.
25. Baughman, R. H., *J. Appl. Phys.* (1972) **43**, 4362.
26. Nakanishi, H., Suzuki Suzuki, F., Hasegawa M., *J. Polymer Sci. A 1*, (1969) **7**, 713.
27. Colson, J. P., Reneker, D. H., *J. Appl. Phys.* (1970) **41**, 4296.
28. Voigt-Martin, I., 23rd IUPAC Congress, Boston, 1971, Macromolecular Preprints Vol. II, p. 818.
29. Kaiser, J., Ph. D. Thesis, Mainz (1972).
30. Hasegawa, M., Suzuki, Y., Tamaki, T., *Bull. Chem. Soc. (Japan)* (1970) **43**, 3020.
31. Takeda, K., Wegner, G., *Makromol. Chem.* (1972) **160**, 349.
32. Morawetz, H., *Science* (1966) **152**, 705.

RECEIVED April 13, 1972.

Photopolymerization of Urethane-Modified Methacrylates for Insulating Magnet Wire

EUGENE D. FEIT

Bell Laboratories, Murray Hill, N. J. 07974

Magnet wire possessing good electrical insulation can be made by photopolymerizing urethane-modified methacrylates. Insulated wire has been produced photochemically in a laboratory-scale reactor at a production rate that can be easily extrapolated to production speeds. Mixtures of the monomers—in this work, monourethane monomethacrylate and diurethane dimethacrylate—are solventless, free-flowing materials with low vapor pressures at room temperature, and they polymerize rapidly in the specially designed apparatus. The films formed exhibit a usable range of mechanical properties and high electrical resistance values.

The Bell System uses millions of pounds of magnet wire annually in the fabrication of electromechanical relays. The need for magnet wire is expected to increase throughout the decade. Before expanding capacity for producing magnet wire, Western Electric requested Bell Laboratories to evaluate new materials and techniques as possible alternatives to the current Bell System practice. It was hoped that a change in material or technique would reduce operating costs, give improved properties, and minimize air pollution in the curing step.

The paper describes a process for the insulation of magnet wire by direct photopolymerization of modified methacrylates. A modified polymethacrylate insulation is attractive for several reasons:

(1) Methacrylates polymerize by a free-radical process that is compatible with high production speeds.

(2) Free-radical polymerization can be initiated by a wide range of energy sources. In principle, methacrylates can be cured not only by thermal activation, but also by ultraviolet and electron radiation.

(3) High-energy radiation permits operation of the coating equipment near ambient temperatures. Low-temperature operation leads to lower vapor losses than at high temperature.

(4) High-energy radiation can incorporate reactive solvent molecules into the final film. Thus, systems in which total polymerization of the applied material occurs are possible.

(5) Polyalkyl methacrylates decompose at temperatures accessible to rapid soldering techniques.

Polymer System

Urethane-modified acrylates and methacrylates are known from the literature (1, 2). These monomers are high boiling and viscous enough to be suitable for solventless applications. Two interesting monomers of this type have been prepared from readily available commercial materials. Compound I is a monourethane monomethacrylate; it is a liquid with a viscosity of 40 cps at 28°C. Compound II is a diurethane dimethacrylate; it is a white, waxy solid that melts around 65°C. Several other monomers of this type have been examined, but these two form the basis of this work.

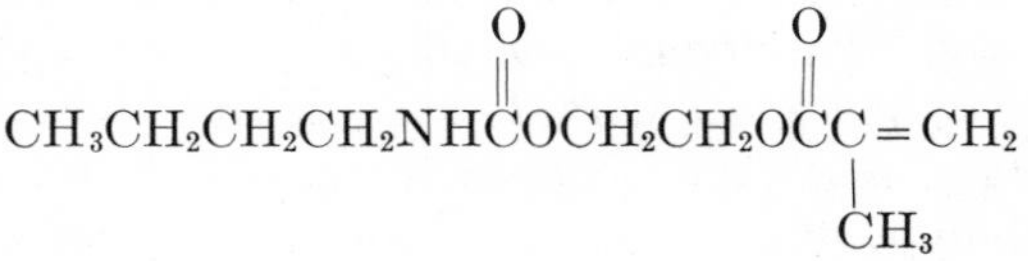

$$CH_3CH_2CH_2CH_2NH\overset{\overset{O}{\|}}{C}OCH_2CH_2O\overset{\overset{O}{\|}}{C}\underset{\underset{CH_3}{|}}{C}=CH_2$$

I

$$(-CH_2CH_2CH_2NH\overset{\overset{O}{\|}}{C}OCH_2CH_2O\overset{\overset{O}{\|}}{C}\underset{\underset{CH_3}{|}}{C}=CH_2)_2$$

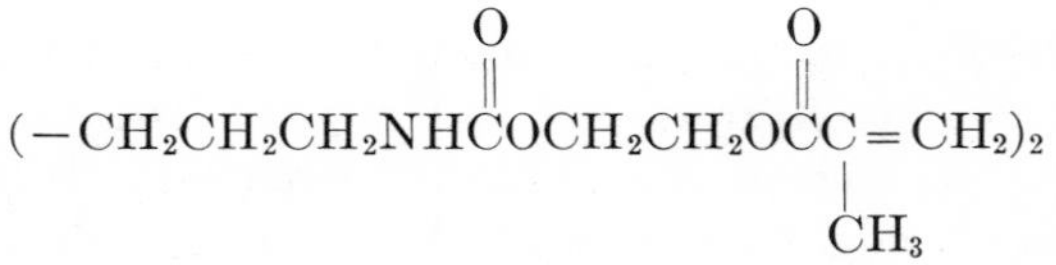

II

Mixtures of I, II, and 1% of benzoin methyl ether (3) (BME) were polymerized with UV radiation of 3.2 mW cm^{-2} intensity. As shown in Figure 1, the infrared absorbance at 6.1μ is reduced by 80% after 15 seconds and is undetectable after 30 seconds of exposure.

A film of a 5:1 (I:II) copolymer is strong (tensile strength = 2700 psi) and 53% extended at break (Table I). A film of a 1:1 copolymer is strong but shows less extensibility than the previous film. The homopolymer of II is excessively brittle.

Flexibility was evaluated by polymerizing the material on 5-mil thick copper foil and subjecting the foil to a ⅛″ mandrel bend. Mixture of I with up to 40% of II shows no cracking of the polymer film after 10 bends.

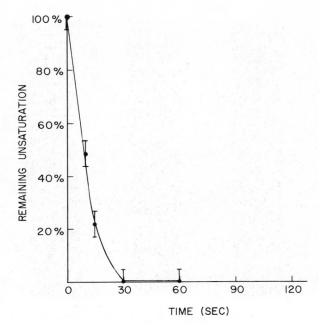

*Figure 1.   Rate of polymerization intensity, 3.2
mW/cm²*

Table I.   Tensile Properties of Urethane Modified Polymethacrylates [a]

1%	2%	Tensile Strength, psi [b]	Elongation, % [b]
82	17	2700	53
50	50	6100	10
0	99	5400	3

[a] Films were formed by UV radiation of the samples (4 mils thick) held between borosilicate glass plates. The photoinitiator was BME at ∼1% by weight.
[b] The entries represent the value at break

These copper strips were soldered for five seconds or less at 400°C. Figure 2 shows the effect of the solder dip on a copolymer of I and II. The results are typical of a range of variations. The strips show that the solder removes the insulation and that there is little shrinkage of the insulation above the solder front.

Insulation resistance measurements of polymer films of these monomers were made on substrates containing interleaved copper-comb patterns. The material was kept at 30°C, 90% humidity, and constant voltage. Insulation resistance values were greater than 10^5 megohms.

In summary, films formed from these materials show a usable range of mechanical properties and high electrical-resistance values. Mixtures

Figure 2. Solder test specimens; two-inch immersion for five seconds.

of the monomers are solventless, free-flowing materials with low vapor pressure at room temperature, and they polymerize at rapid rates.

Apparatus

The evaluation of these materials as wire enamels must be made on the basis of tests performed on insulated wire. A laboratory enameling machine has been built that can produce hundreds of feet of enameled wire suitable for testing.

The principle components of the apparatus are shown in Figure 3. Briefly, the wire pays-off into an applicator and then into a vertical photoreactor. The drive and take-up assemblies are situated behind the reactor and are designed to minimize tension on the wire. The photoreactor is the core of the apparatus and is discussed extensively here.

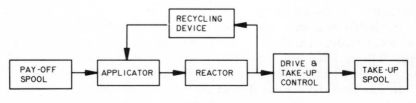

Figure 3. Schematic of apparatus

The reactor consists basically of two hemicylinders hinged together on a vertical side. The wire enters the reactor through a small aperture in the base and leaves the reactor through a similar aperture in the top. Six low-pressure mercury lamps (Southern New England Ultraviolet Co.) are mounted vertically and symmetrically inside the reactor. The principal emission (~80%) from these lamps occurs at 254 nm. A principal emission at 300 or 350 nm is available from lamps constructed with the appropriate phosphors (Figure 4). The inside walls of the reactor are made reflecting to UV radiation by flexible, front-sided mirrors. The

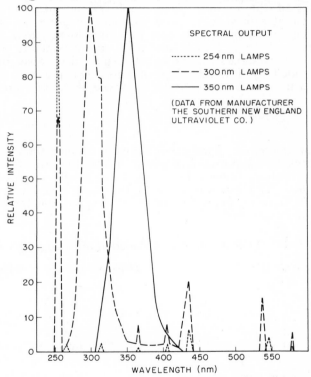

Figure 4.   Spectral output of lamps used in the photo-reactor.

mirrors reflect 75-85% of the radiation between 220-300 nm as produced (Denton Vacuum) and, after 100 hours of use, they show little loss of reflectance. The intensity of radiation in the reactor was measured with a calibrated Eppley thermopile. In the open reactor, the radiation intensity from three lamps with 254 nm principal emission was 2.0 ± 0.5 mW/cm^2 along 90% of the vertical axis. In the closed reactor, the intensity at the same positions from six lamps rose to 4-5 mW/cm^2. The values of these last measurements are affected somewhat by the shadow cast by the large mass of the thermopile. The UV intensities from the lamps with principle emissions at 300 and 350 nm are similar to the intensity from the 254-nm lamps. This result is not surprising in view of the efficiency of the phosphors. (4).

The atmosphere in the reactor can be controlled by inlets at top and bottom. The reactor is sealed with a flexible, self-forming gasket. The temperature inside the closed reactor with the 254-nm lamps operating and with nitrogen flowing at 0.5 cubic feet per minute was stable at 72°C. This temperature rise is expected to increase slightly the lamp intensities recorded in the preceding paragraph (5). The wire is routed inside the reactor by a set of pulleys mounted on the top and bottom inside plates of the reactor. This simple arrangement increases the exposure time at a fixed wire speed without increasing the size of the reactor. The applicator consists of a tank to hold the resin, a grooved wheel to guide the wire through the resin, and a wiper to smooth the resin before it polymerizes. Friction between the drive assembly and the wire is provided by a weighted pivot arm. The pivot of the arm is coupled to a transducer that controls the speed of the take-up assembly.

Operation

In actual operation, wire pays-off into a furnace for cleaning and annealing and then into a tank of resin. The resin is transferred to the wire by dip coating, smoothed, and photochemically polymerized in a sealed reactor. The exposure time in the reactor can vary from 10 seconds to 52 minutes, depending on capstan speed and the number of pulleys inside the reactor.

Table II.   Defect Count as a Function of Nitrogen Flow

Flow Rate (CFM) [a]	Defect Count (per 100 feet)
0.50 [b]	0,0 [c]
0.43	0,0,0
0.36	0,0,0
0.30	29,24
0.25	463,424

[a] Inlet pressure is 2 psi of nitrogen
[b] Partial pressure of oxygen ~2 torr
[c] Entries are the number of discontinuities in insulation counted per 100 feet of continuously produced and tested magnet wire.

Once it has left the reactor, the wire may be either collected directly on the take-up *via* the drive assembly or returned to the resin tank for application of another layer of insulation *via* the pulley in Figure 3. Three layers of insulation are usually applied to the wire before take-up. Typical formulations for the experiments contain about 1% of a processing aid—*e.g.*, Nuosperse 657 (Nuodex Products Co.). Without the processing aid, the formulations wet copper poorly and bead on the wire. The processing aid is used to smooth the resin and furnish a uniform thickness of insulation. Surface contaminants on the conductor affect the continuity of the insulation. A forced, hot-air furnace serves as a simple method of pretreatment. The resin polymerizes in the photochemical reactor. A flow of pure nitrogen (< 10 ppm of oxygen) into the reactor is necessary because the polymerization is inhibited by oxygen.

Variations in operational parameters are judged by their effect on the continuity of insulation on the wire. A continuity test—sometimes called a "pinhole" test—is an ASTM test (6) performed on commercially available equipment (Hypotronic, Inc.). In the test, a defect in the insulation is counted each time current flows between a 1-inch section of the insulated wire and a mercury bath through which the wire travels. A time constant (25 msec) is associated with the recovery of the counting device in such a way that a single, prolonged defect is counted many times as a series of shorter (0.5-inch) defects. The dependence of defect count on flow rate for continuous production of insulated wire is given in Table II. "Defect-free" wire is produced at 4 ft/min at a partial pressure of oxygen ∼ 2 torr. This partial pressure is obtained at a nitrogen flow of 0.5 cubic feet per minute. The flow rate required depends on the design of the reactor. The partial pressure of oxygen is a fundamental parameter.

Table III.   Defect Count as a Function of Wire Speed

Speed (ft/min)	Exposure (seconds)	Defect Count (per 100 feet)
4.0	45	0,3 [a]
4.5	40	0,0,1
5.0	36	3,1,0,0
5.5	33	0,3,1,0,0
6.0	30	29,14,15

[a] Entries are the number of discontinuities in insulation counted per 100 feet of continuously produced and tested magnet wire.

A residence time in the reactor of 12-15 seconds sufficiently hardens the resin to allow rerouting the coated wire inside the reactor for longer, total exposure. The dependence of defect count on wire speed and total exposure is shown in Table III.

Production of Insulated Wire

Twelve thousand feet of wire were continuously insulated in the apparatus at 4 ft/min with three layers of insulation. Defect counts were recorded for every 50-foot section of wire. The frequency of occurrence of a count is plotted against the count in Figure 5. Defect-free, 50-foot sections predominate. The average defect count per 100 feet is 0.6.

An additional 16,000 feet of wire were insulated, and 1600 feet were randomly selected for testing. No defects were counted, indicating that defects when they do occur tend to occur in clusters and at the start of the operation. This observation is consistent with data from the first 12,000 feet. D. C. breakdown values for the insulated wire were measured as 2.52, 2.22, 0.88, 2.00 and 2.36 kV (6).

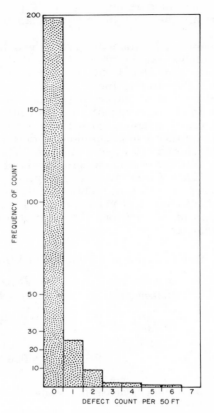

Figure 5.   Frequency of defect counts

Summary

In summary, then, magnet wire with good and continuous electrical insulation can be produced by photopolymerization of urethane-modified methacrylates. Insulated wire is produced photochemically in the laboratory scale reactor at 4 ft/min. This production rate can easily be extrapolated to production speeds. The length of the reactor can be extended sixfold. Radiation intensity can be increased ten times or more with high-pressure arcs. If production rates are proportional even to the half power of the intensity, the rate will be tripled or more. Operating at a higher ambient temperature in the reactor and optimizing the photochemical sensitivity of the resin can double or triple improvement in rate. A final rate of production of 150-200 ft/min, then, is within the range of the final process with this resin.

Acknowledgment

The author acknowledges W. Bader, J. Lynch, J. MacKay, T. F. Halloran, and L. D. Loan of Bell Laboratories, and P. R. McVickers and M. A. MacVittie of Western Electric, Buffalo.

Literature Cited

1. Nordstrom, J. D., U.S. Patent **3,479,328** (Nov. 18, 1969).
2. Labana, S. S., *J. Polym. Sci., A-1* (1968) **6**, 3283.
3. Kosar, J., "Light Sensitive Systems," pp. 162-3, Wiley, New York, 1965.
4. Koller, L. R., "Ultraviolet Radiation," 2nd edition, p. 264, Wiley, New York, 1965.
5. *Ibid.*, page 52.
6. ASTM D1676-70.

RECEIVED August 2, 1972.

INDEX

INDEX

The text of this book is set in 10 point Caledonia with two points of leading. The chapter numerals are set in 30 point Garamond; the chapter titles are set in 18 point Garamond Bold.

The book is printed offset on Danforth 550 Machine Blue White text, 50-pound. The cover is Joanna Book Binding blue linen.

Jacket design by Norman Favin.
Editing and production by Mary Westerfeld.

The book was composed by Modern Linotypers, Inc., Baltimore, Md., printed and bound by The Maple Press Co., York, Pa.